Injection-Locking in Mixed-Mode Signal Processing

Fei Yuan

Injection-Locking in Mixed-Mode Signal Processing

Springer

Fei Yuan
Electrical and Computer Engineering
Ryerson University
Toronto, ON, Canada

ISBN 978-3-030-17362-3 ISBN 978-3-030-17364-7 (eBook)
https://doi.org/10.1007/978-3-030-17364-7

This Springer imprint is published by the registered company Springer Nature Switzerland AG.
The registered company address is: Gewerbestrasse 11, 6330 Cham, Switzerland

Preface

The injection-locking or synchronization of an oscillator by an external periodic signal hereafter referred to the injection signal whose frequency is in the proximity of the natural frequency, more specifically the frequency of the first harmonic, of the oscillator under injection shifts the frequency of the oscillator from its natural frequency to the frequency of the injection signal. The frequency of the injection-locking signal can also be either a superharmonic or a subharmonic of the natural frequency of the oscillator under injection to realize frequency division or frequency multiplication. First-harmonic injection-locking achieves frequency synchronization by utilizing the frequency dependence of the amplitude and phase of the output of the oscillators under injection, whereas superharmonic/subharmonic injection-locking achieves frequency-locking by utilizing the internal nonlinear mechanism of the oscillator to generate the desired frequency component and the internal frequency selection mechanism of the oscillator to select the desired frequency component generated by the nonlinear mechanism of the oscillator. The absence of an explicit frequency detector, a charge pump, and a loop filter that typically exist in a frequency-locked loop (FLL) greatly shortens frequency-locking process and lowers power consumption. Injection-locking also provides a better frequency accuracy as compared with conventional FLLs. Not only the frequency of the oscillator under injection is shifted to that of the injection signal, the phase noise of the oscillator under injection will also become comparable to that of the injection signal once the oscillator is locked to the injection signal whose phase noise is superior, thereby allowing low-cost oscillators whose phase noise is typically inferior to function as oscillators with superior phase noise performance.

Injection-locking has found a broad range of emerging applications including the pre-scalar of frequency synthesizers, forwarded-clock (FC) or source synchronous parallel links, the frequency calibration of passive wireless microsytems (PWMs), and low phase noise phase-locked loops, to name a few. The success of the injection-locking of an oscillator critically depends upon the lock range of the oscillator, which is the range of the difference between the frequency of the injection signal and the natural frequency of the oscillator in which the establishment of the lock state of the oscillator under injection is guaranteed. The lock range of an oscillator is not only determined by the characteristics of the oscillator but also by the characteristics of the injection-locking signal. Many novel design techniques to increase the lock range of oscillators and analysis methods to quantify the characteristics of injection-locking of oscillators have emerged since the inception of the synchronization of oscillators using external signals approximately a century ago. A comprehensive treatment of both the fundamentals of the injection-locking of oscillators and an in-depth assessment of the pros and cons of the techniques that have been proposed to increase the lock range of injection-locked oscillators are not available. This monograph is an attempt to provide a comprehensive treatment of the principles and design techniques of injection-locking in mixed-mode signal processing.

The book is organized as follows:

Chapter 1 provides an overview of injection-locking and its applications in mixed-mode signal processing. The classification of oscillators is provided. It browses through the development of the injection-locking of oscillators with an emphasis on the characterization rather than the design of injection-locked oscillators. First-harmonic methods for analyzing harmonic oscillators in weak injection are presented. It is followed with the presentation of first-harmonic methods for the analysis of harmonic oscillators in both weak and strong injection. Frequency regenerative injection specifically tailored for frequency multiplication and frequency division is explored. First-harmonic balance method capable of analyzing harmonic oscillators in first-harmonic, superharmonic, and subharmonic injections is studied. The progressive multiphase injection of ring oscillators with multiple injections is examined. The effective injection signaling arising from the nonlinearity of oscillators under injection and obtained by analyzing the Volterra circuits of the oscillators under injection is described. The chapter also briefly browses through the key representative applications of the injection-locking of oscillators.

Chapter 2 presents the fundamentals of the oscillation of harmonic oscillators first. It is followed with a close examination of the noise spectrum of harmonic oscillators. The modeling of injection-locked harmonic oscillators using a nonlinear system and the first-harmonic injection-locking of harmonic oscillators are investigated. Both linear and nonlinear approaches capable of deriving the lock range of harmonic oscillators are studied. The chapter also investigates the superharmonic injection-locking of harmonic oscillators. Both the second-order and third-order superharmonic injection-locking of harmonic oscillators are studied. Divide-by-2 and divide-by-3 injection-locked frequency dividers, which are the representative applications of the superharmonic injection-locking of harmonic oscillators, are studied in detail. The subharmonic injection-locking of harmonic oscillators is investigated. The intrinsic relations among the lock range of harmonic oscillators in first-harmonic, superharmonic, and subharmonic injection-locking are explored. Finally, the phase noise of injection-locked harmonic oscillators is studied.

Chapter 3 explores the factors that affect the lock range of harmonic oscillators and the techniques that increase the lock range of harmonic oscillators. Our focus is on injection-locked frequency dividers where injection-locked harmonic oscillators are mostly encountered. The chapter further investigates the dependence of the lock range of harmonic oscillators on injection signaling. Specifically, we investigate the lock range of harmonic oscillators with two unitone injections injected into two symmetrical locations of the oscillators and the relation between the lock range and the phase of the injection signals. Various techniques to increase the lock range of harmonic oscillators including inductor series-peaking, transformer series-peaking, inductor shunt-peaking, current-reuse direct injection, quality factor reduction, resistor feedback, second harmonic extraction, transformer feedback, and dual injections are studied.

Chapter 4 investigates the oscillation of ring oscillators, in particular, how Barkhausen criteria can be used to analyze ring oscillators. The modeling of harmonic oscillators is investigated. The modeling of relaxation oscillators is explored with a special attention to the distinct nonlinear characteristics of these oscillators as compared with ring oscillators. The representation of a nonharmonic oscillator with a set of harmonic oscillators whose frequencies are harmonically related so as to allow linear system analysis tools such as Laplace transform valid for analysis of harmonic oscillators and the known knowledge of harmonic oscillators to be fully utilized in analysis of nonharmonic oscillators. The fundamentals of Volterra series are reviewed. Two representative nonlinearities widely encountered in integrated circuits, specifically the channel current of MOS transistors and the capacitance of pn-junctions, are studied. The concept of the Volterra elements of a nonlinear element and the Volterra circuits of a nonlinear circuit are introduced, and the process of how to obtain them is exemplified. The modeling of voltage comparators is studied. We show that the nonlinear voltage transfer characteristics of comparators can be depicted using a hyperbolic tangent function and the hyperbolic tangent function can be approximated using its third-order Taylor series expansion in the vicinity of the state transition of the comparator. The Volterra circuits of an injection-locked nonharmonic oscillator are derived, and the characteristics of the Volterra circuits are investigated. The chapter explores how the Volterra circuit approach can be used to analyze the dual-comparator relaxation oscillator under the injection of a pair of differential currents and how the high-order Volterra circuits of the oscillator contribute to the effective injection signals of the first-order Volterra circuit of the oscillator. Finally, the lock range of the dual-comparator relaxation oscillator is investigated.

Chapter 5 deals with the techniques that increase the lock range of nonharmonic oscillators. It studies the impact of the degree of the nonlinear characteristics of nonharmonic oscillators on the lock range of these oscillators and investigates the lock range of dual-comparator relaxation oscillators with single unitone injection and that with dual unitone injections. The lock range of dual-comparator relaxation oscillators with dual multitone injections is derived. It compares the lock range of dual-comparator relaxation oscillators with single multitone injection and that with dual multitone injections. The impact of the duty cycle of multitone injection signals on the lock range of nonharmonic oscillators is investigated. The impact of the frequency attenuation mechanism on the lock range of both harmonic and nonharmonic oscillators is investigated and compared.

The monograph is intended to cover the state of the art of injection-locking in mixed-mode signal processing. Readers are assumed to have the fundamental knowledge of electric networks, semiconductor devices, microelectronic circuits, signals and systems, analog and digital communication systems, and analog and digital integrated circuits and systems. A rich collection of recently published work on the injection-locking of oscillators is provided at the end of the monograph for readers to seek further information on the subjects that are presented in the monograph.

Although an immense effort was made in preparation of the manuscript, flaws and errors will surely still exist due to both erring human nature and the limited knowledge of the author on the presented subjects. Suggestions and corrections from readers will be gratefully appreciated by the author.

Toronto, ON, Canada Fei Yuan
February 2019

Acknowledgments

I am grateful to the Natural Sciences and Engineering Research Council of Canada; CMC Microsystems, Kingston, ON, Canada; Ryerson University; and other research partners for their financial support to our research. The support from the Department of Electrical, Computer, and Biomedical Engineering of Ryerson University, Toronto, ON, Canada, where I introduced and taught a number of senior undergraduate and graduate courses on CMOS mixed-mode circuits, CMOS circuits and systems for data communications over wire channels, and CMOS radio-frequency circuits and systems is gratefully acknowledged. Special thanks go to my former graduate students who worked with me on various research projects in the field of the injection-locking of oscillators from which many of the original contributions presented in this monograph emerged. The recently granted sabbatical leave by the university gave me much needed time release to complete this long overdue project.

The editorial staff of Springer, especially Mr. Charles Glaser, Editorial Director for Electrical and Computer Engineering: Circuits and Systems, Design Automation, Embedded Systems, and Computer Architecture, and Ms. Brinda Megasyamalan, Project Coordinator and Production Editor, have been warmly supportive from the submission of the proposal to the publishing of the book. This is the fifth book that I have published with Springer. It has been an enjoyable experience of working with Springer.

Finally and most importantly, this book could not have been possible without the unconditional support of my family. I am indebted to my wife Jing for her love, patient, and understanding during the preparation of the manuscript and her tolerance of my occasional bad temper due to the stress of the writing and proofreading of the manuscript.

Contents

About the Author

Fei Yuan received the B.Eng. degree in electrical engineering from Shandong University, Jinan, China, in 1985 and the M.A.Sc. degree in chemical engineering and the Ph.D. degree in electrical engineering from the University of Waterloo, ON, Canada, in 1995 and 1999, respectively. During 1985–1989, he was a Lecturer in the Department of Electrical Engineering, Changzhou Institute of Technology, Jiangsu, China. In 1989, he was a Visiting Professor at Humber College of Applied Arts and Technology, Toronto, ON, Canada, and Lambton College of Applied Arts and Technology, Sarnia, ON, Canada. He was in Paton Controls, Sarnia, ON, Canada, during 1989–1994 where he designed distributed process control systems for petrochemical processes worldwide. Since 1999, he has been with the Department of Electrical, Computer, and Biomedical Engineering, Ryerson University, ON, Canada, where he is currently a Full Professor. He served as the Chair of the Department during 2010–2015.

Dr. Yuan is the Editor of *Low-Power Circuits for Emerging Applications in Communications, Computing, and Sensing* (CRC, 2018); the Editor and a lead Contributor of *CMOS Time-Mode Circuits: Principles and Applications* (CRC, 2015); the Author of *CMOS Circuits for Passive Wireless Microsystems* (Springer, 2010), *CMOS Active Inductors and Transformers: Principle, Implementation, and Applications* (Springer, 2008), and *CMOS Current-Mode Circuits for Data Communications* (Springer, 2006); and the lead Coauthor of *Computer Methods for Analysis of Mixed-Mode Switching Circuits* (Kluwer Academic, 2004). He is the Author/Coauthor of 10 book chapters and over 230 research papers in refereed international journals and conference proceedings. He was awarded Dean's Teaching Award in 2017, Ryerson Research Chair Award in 2005, Dean's Research Award in 2004, Early Tenure from Ryerson University in 2003, Doctoral Scholarship from the Natural Sciences and Engineering. Research Council of Canada in 1997 and 1998, Teaching Excellence Award from Changzhou Institute of Tech. in 1988, and Science and Technology Innovation Award from Changzhou Municipal government in 1988. He is a Fellow of IET, a Senior Member of IEEE, and a Registered Professional Engineer in the province of Ontario, Canada.

Symbols

C_{ox}	Gate capacitance per unit area of MOS transistors
C_{gs}	Gate-source capacitance of MOS transistors
C_{gd}	Gate-drain capacitance of MOS transistors
C_J	Capacitance of an pn-junction
g_m	Transconductance of MOS transistors
$g_{m,n}$	Transconductance of NMOS transistors
$g_{m,p}$	Transconductance of PMOS transistors
g_{ds}	Channel conductance of MOS transistors in triode
g_o	Output conductance of MOS transistors in saturation
k	Boltzmann's constant
K_f	Flicker noise coefficient of MOS transistors
H_o	Open-loop transfer function
H_c	Closed-loop transfer function
$H(\omega_{out}, \omega_{in})$	Aliasing transfer function
i_{DS}	Channel current of NMOS transistors
I_{DS}	Channel current of NMOS transistors (DC)
i_{ds}	Channel current of NMOS transistors (AC)
i_{SD}	Channel current of PMOS transistors
I_{SD}	Channel current of PMOS transistors (DC)
i_{sd}	Channel current of PMOS transistors (AC)
L	Channel length of MOS transistors
r_{ds}	Channel resistance of MOS transistors in triode
r_o	Output resistance of MOS transistors in saturation
Q	Quality factor
S_{in}	Power spectral density of input
S_{out}	Power spectral density of output
s	Laplace variable ($s = j\omega$)
V_{Tn}	Threshold voltage of NMOS transistors
V_{Tp}	Threshold voltage of PMOS transistors
v_{GS}	Gate-source voltage of NMOS transistors
V_{GS}	Gate-source voltage of NMOS transistors (DC)

v_{gs}	Gate-source voltage of NMOS transistors (AC)
v_{SG}	Source-gate voltage of PMOS transistors
V_{SG}	Source-gate voltage of PMOS transistors (DC)
v_{sg}	Source-gate voltage of PMOS transistors (AC)
v_{DS}	Drain-source voltage of NMOS transistors
V_{DS}	Drain-source voltage of NMOS transistors (DC)
v_{ds}	Drain-source voltage of NMOS transistors (AC)
v_{SD}	Source-drain voltage of PMOS transistors
V_{SD}	Source-drain voltage of PMOS transistors (DC)
v_{sd}	Source-drain voltage of PMOS transistors (AC)
W	Channel width of MOS transistor
μ_n	Surface mobility of free electrons
μ_p	Surface mobility of holes
γ	Thermal noise coefficient of MOS transistors
$\Delta\omega_L$	Lock range of injection-locked oscillator
ω_o	Natural or free-running frequency of an oscillator
ω_{inj}	Frequency of an injection signal
ϕ_{inj}	Phase of an injection signal
ϵ	Perturbation constant for the use of Volterra series

Acronyms

ASK	Amplitude-Shift Keying
BER	Bit Error Rate
CDR	Clock and Data Recovery
CMOS	Complementary Metal-Oxide Semiconductor
CML	Current-Mode Logic
DAC	Digital-to-Analog Converter
dB	Decibel
DC	Direct Current
DFF	D Flip-Flop
DLL	Delay-Locked Loop
EPC	Electronic Product Code
FC	Forwarded-Clock
FET	Field-Effect Transistor
FF	Flip-Flop
FFT	Fast Fourier Transform
FIR	Finite Impulse Response
FLL	Frequency-Locked Loop
Gbps	Gigabits per second
GHz	Gigahertz
High-Q	High-Quality factor
Hz	Hertz
IBM	International Business Machines Corporation
IC	Integrated Circuit
IF	Intermediate Frequency
IIR	Infinite Impulse Response
ILFD	Injection-Locked Frequency Divider
IoT	Internet-of-Things
kHz	Kilohertz
KCL	Kirchhoff's Current Law
KVL	Kirchhoff's Voltage Law
LC	Inductor Capacitor

MDLL	Multiplying Delay-Locked Loop
MHz	Megahertz
MIM	Metal-Insulator-Metal
MOS	Metal-Oxide Semiconductor
MOSFET	Metal-Oxide Semiconductor Field-Effect Transistor
NMOS	N-type Metal-Oxide Semiconductor
OPAMP	Operational Amplifier
PFD	Phase-Frequency Detector
PI	Phase Interpolator
PLL	Phase-Locked Loop
PMOS	P-type Metal-Oxide Semiconductor
PSD	Power Spectral Density
PVT	Process, Voltage, and Temperature
PWM	Passive Wireless Microsystem
RC	Resistor Capacitor
RF	Radio Frequency
RFID	Radio Frequency Identification
RL	Resistor Inductor
RLC	Resistor Inductor Capacitor
UHF	Ultra High Frequency
UI	Unit Interval
V2I	Voltage-to-Current
SR	Set-Reset
SPICE	Simulation Program with Integrated Circuit Emphasis
VCO	Voltage-Controlled Oscillator

Chapter 1
Injection-Locking of Oscillators: An Overview

1.1 Introduction

The injection-locking or synchronization of an electronic oscillator by an external periodic signal hereafter referred to as the injection signal whose frequency is in the proximity of the natural or free-running frequency, more specifically the frequency of the first harmonic, of the oscillator under injection shifts the frequency of the oscillator from its natural frequency to the frequency of the injection signal without a frequency-locked loop, which is more costly in terms of both silicon area and power consumption. First-harmonic injection-locking achieves frequency synchronization by utilizing the frequency dependence of the amplitude and phase of the output of the oscillators under injection, whereas superharmonic/subharmonic injection-locking achieves frequency-locking by utilizing the internal nonlinear mechanism of the oscillator under injection that functions as a frequency mixer to generate the desired frequency component and the internal frequency selection mechanism of the oscillator to select the desired frequency component generated by the nonlinear mechanism of the oscillator. In both cases, Barkhausen criteria for oscillation must be satisfied. The absence of an explicit frequency detector, a charge pump, and a loop filter that typically exist in a FLL not only greatly shortens the frequency-locking process, but also significantly lowers power consumption. Injection-locking is capable of achieving rapid frequency-locking pivotal to a broad range of applications. Not only the frequency of the oscillator under injection is shifted to that of the injection signal, the phase noise of the oscillator under injection will also become comparable to that of the injection signal once the oscillator is locked to the injection signal whose phase noise performance is superior. This unique characteristic of injection-locked oscillators allows low-cost oscillators whose phase noise is inferior typically to behave as oscillators with superior phase noise performance once the oscillators are locked to a clean signal.

This chapter provides an overview of the injection-locking of oscillators and its applications in mixed-mode signal processing. The classification of oscillators is

F. Yuan, *Injection-Locking in Mixed-Mode Signal Processing*,
https://doi.org/10.1007/978-3-030-17364-7_1

provided in Sect. 1.2. Section 1.3 briefly browses through the development of the synchronization of oscillators prior to 1946 in which the ground-breaking work of Adler on the injection-locking of harmonic oscillators was published. Section 1.4 presents Adler's first-harmonic method for analysis of harmonic oscillators in weak injection. It is followed with the presentation of the first-harmonic method for analysis of harmonic oscillators in strong injection in Sect. 1.5. Frequency regenerative injection specifically tailored for frequency multiplication and frequency division is explored in Sect. 1.6. Section 1.7 studies the first-harmonic balance method capable of analyzing harmonic oscillators in first-harmonic, superharmonic, and subharmonic injections. The progressive multiphase injection of ring oscillators with multiple injections is examined in Sect. 1.8. Further development of this method for multiphase harmonic oscillators and injection-locked ring oscillator frequency dividers is also referenced. Section 1.9 briefly presents a Volterra series-based approach to investigate the impact of the degree of the nonlinearity of an oscillator under injection on the lock range of the oscillator. Section 1.10 presents some key applications of the injection-locking of oscillators. The chapter is summarized in Sect. 1.11.

1.2 Classification of Oscillators

Oscillators can be loosely classified into harmonic oscillators and nonharmonic oscillators. The former have a sinusoidal output in the time domain and a unitone spectrum in the frequency domain, whereas the latter have a non-sinusoidal output in the time domain and a multitone spectrum in the frequency domain. The typical examples of harmonic oscillators are LC (inductor-capacitor) oscillators with either a spiral inductor/transformer resonator or an active inductor/transformer resonator. The representative nonharmonic oscillators include ring oscillators and relaxation oscillators.

Architecturally a harmonic oscillator is made of two distinct blocks, namely (a) a nonlinear transconductor that functions as an amplitude limiter needed to sustain oscillation and a frequency mixer where various frequency components are generated from the injection signal and the output of the oscillator when the oscillator is under the injection of an external signal and (b) a linear frequency discriminator that functions as a high-Q band-pass filter to select the desired output frequency. Figure 1.1a shows the simplified block diagram of a harmonic oscillator under the injection of unitone external signal v_{inj}. Similarly, a nonharmonic oscillator is composed of a nonlinear transconductor that functions as both an amplitude limiter and a frequency mixer when the oscillator is under injection of an external signal and a linear RC block that functions as a low-pass filter, as shown in Fig. 1.1b.

Harmonic oscillators distinct themselves from nonharmonic oscillators with a small frequency tuning range and low phase noise, accredited mainly to the higher quality factor of the frequency discriminator of the oscillators. The superior phase

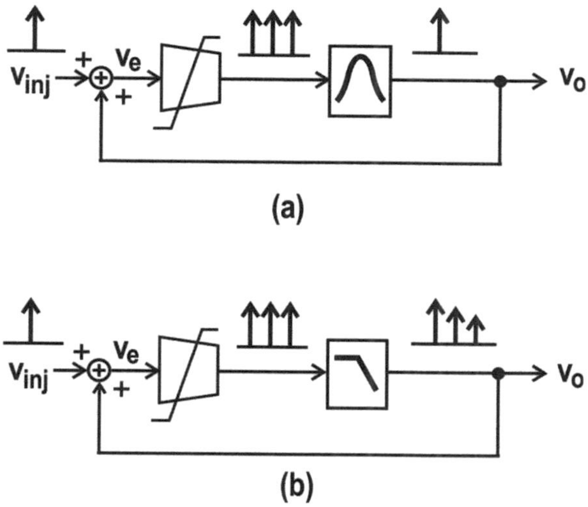

Fig. 1.1 Block diagram of oscillators under injection: (**a**) harmonic oscillators. (**b**) Nonharmonic oscillators

noise performance of harmonic oscillators allows them to be used in wireless communication systems where a stringent constraint on the phase noise of oscillators exists. Nonharmonic oscillators, on the other hand, are advantageous over their harmonic counterparts in silicon area, power consumption, and frequency tuning range. They are widely used in digital systems as clock generators.

1.3 Injection-Locking of Oscillator: Before 1946

The earliest observation of the synchronized oscillation of two oscillating systems is perhaps the synchronized pendulums of two pendulum clocks by C. Huygens (1629–1695), a Dutch physicist and the inventor of the pendulum clock, in 1665. Huygens noticed that no matter how the pendulums on these clocks began, within about half an hour, they ended up swinging in exactly the opposite directions from each other. Perhaps the earliest reported work on the synchronization of oscillators are those by J. Vincent in 1919 [84] and E. Appleton in 1922 [2]. One of the first theoretical analyses of the nonlinear characteristics of oscillators was provided by van der Pol in 1934 [81]. An early application of modulating the frequency of an electronic oscillator by an external signal is the frequency-modulation receiver by G. Beers in 1944 where the frequency of the oscillator of the receiver is modulated in accordance with the frequency of an external modulating signal [6]. An in-depth study of forced oscillation in oscillator circuits was conducted by D. Tucker in 1945 [79, 80]. In Tucker's approach, an oscillator was partitioned into two functionally distinct blocks, specifically a frequency discriminating block that functions as a band-pass filter and a maintaining block whose transfer voltage ratio is only dependent of the amplitude of the input and independent of the frequency of the input of the block. Using the phasor representation of the injection signal, the behavior of forced oscillators was analyzed.

1.4 Adler: Weak First-Harmonic Injection

In 1946, R. Adler, an American physicist (1913–2007), published his widely cited theoretical work on locking phenomena in oscillators [1]. Adler's approach was developed for harmonic oscillators under the injection of a weak continuous-wave injection signal. The following three constraints were imposed:

(a) The frequency of the injection signal is in the proximity of the natural also known as free-running frequency of the oscillator under injection.
(b) The strength of the injection signal is significantly smaller as compared with the output of the oscillator, i.e., weak injection.
(c) The feedback time from the output of the oscillator to the node at which the output voltage of the oscillator is subtracted from the injection signal is negligible as compared with beat period $1/(\omega_{inj} - \omega_o)$ where ω_{inj} and ω_o are the frequency of the injection signal and the natural frequency of the oscillator under injection, respectively.

Although the derivation of Adler's theoretical results was rigorous and correct, no experimental validation of the theoretical results was provided in the paper.

Given the significance of Adler's work on the injection-locking of oscillators and its profound impact on the investigation of the injection-locking of oscillators, a brief representation of Adler's approach on the injection-locking of oscillators is clearly warranted and well justified. In what follows we use the simple LC oscillator shown in Fig. 1.2 to briefly present Adler's approach on analysis of the injection-locking of oscillators. As lock range is of our primary interest, we will focus on the lock range of the oscillator.

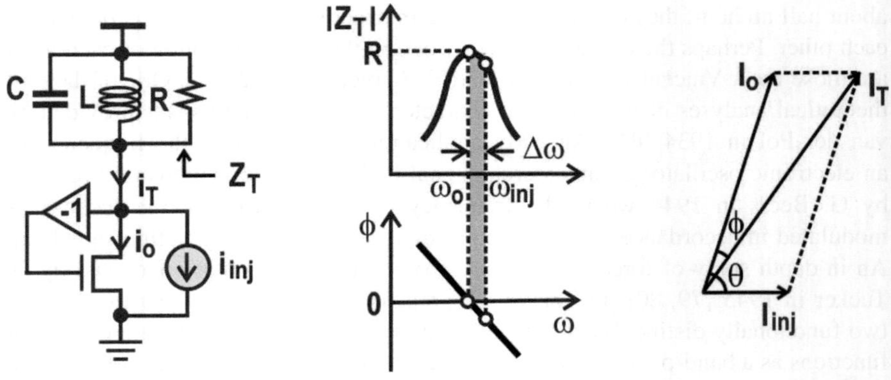

Fig. 1.2 Injection-locking in harmonic oscillators

Consider the simplified schematic of the LC oscillator shown in Fig. 1.2. It is a tuned common-source amplifier with the addition of an inverting amplifier of unity gain. Since the RLC network functions as a pure resistor at its resonant frequency $\omega_o = 1/\sqrt{LC}$, the common-source tuned amplifier only provides a phase shift of -180° at ω_o. An additional -180° phase shift needed for oscillation is provided by the unity-gain inverting amplifier. When the injection signal of frequency ω_{inj} is injected into the oscillator with ω_{inj} in the close proximity of ω_o, the frequency of the oscillator will start to shift from its natural frequency ω_o to ω_{inj}. Let the instantaneous frequency of the oscillator during the locking process be ω. The instantaneous variation of the frequency also known as the beat frequency of the oscillator is given by $\omega - \omega_o$. Further let θ be the phase between the injection signal I_{inj} and the output of the oscillator I_o. Since when I_{inj} is absent, I_T and I_o will be in phase, I_{inj} is therefore the cause of both frequency shift $\Delta\omega$ and phase angle ϕ between I_T and I_o. Referring to Fig. 1.2, the law of sine gives

$$\frac{I_{inj}}{\sin\phi} = \frac{I_o}{\sin\theta} \tag{1.1}$$

from which we obtain

$$\sin\phi = \frac{I_{inj}}{I_o}\sin\theta. \tag{1.2}$$

If $I_{inj} \ll I_o$, i.e., weak injection, ϕ will be small and

$$\sin\phi \approx \phi \tag{1.3}$$

will hold. Equation (1.2) can be simplified to

$$\phi \approx \frac{I_{inj}}{I_o}\sin\theta. \tag{1.4}$$

As mentioned earlier, the injection signal causes the frequency of the oscillator to deviate from its natural frequency ω_o and gives rise to a phase angle between I_T and I_o. Since the frequency of the injection signal is in the proximity of ω_o, the rate of the variation of ϕ with respect to frequency

$$A = \frac{d\phi}{d\omega} \tag{1.5}$$

can be considered to be constant. We therefore have

$$\begin{aligned}\phi &\approx A(\omega - \omega_o)\\ &= A\left(\Delta\omega + \Delta\omega_L\right).\end{aligned} \tag{1.6}$$

where

$$\Delta\omega = \omega - \omega_{inj} \tag{1.7}$$

is the instantaneous beat frequency and

$$\Delta\omega_L = \omega_{inj} - \omega_o \tag{1.8}$$

is the undisturbed beat frequency. Equating (1.4) and (1.6) and noting that

$$\omega = \frac{d\theta}{dt} + \omega_{inj}, \tag{1.9}$$

we arrive at

$$\frac{d\theta}{dt} + \Delta\omega_L = \frac{1}{A}\frac{I_{inj}}{I_o}\sin\theta. \tag{1.10}$$

The impedance of the parallel RLC network is given by

$$Z(s) = \frac{1}{C}\frac{s}{s^2 + s\frac{1}{RC} + \frac{1}{LC}}. \tag{1.11}$$

It can be shown that the phase of $Z(j\omega)$, denoted by $\angle Z(j\omega)$, is given by

$$\angle Z(j\omega) = -\tan^{-1}\left(\frac{R}{\omega L}\frac{\omega_o^2 - \omega^2}{\omega_o^2}\right). \tag{1.12}$$

When the injection signal is absent,

$$I_o = g_m Z_T(j\omega_o) I_T, \tag{1.13}$$

where g_m is the transconductance of the transistor. Note $Z(j\omega_o) = R$. I_o and I_T are in phase in this case. When the injection current I_{inj} is present,

$$I_o = g_m Z_T(j\omega_{inj}) I_T. \tag{1.14}$$

I_o and I_T in this case are not in phase but rather have a phase angle $\angle Z(j\omega_{inj})$, which is ϕ. We therefore have

$$\phi = -\tan^{-1}\left(\frac{R}{\omega L}\frac{\omega_o^2 - \omega^2}{\omega_o^2}\right). \tag{1.15}$$

Since ω is in the vicinity of ω_o,

$$\omega_o^2 - \omega^2 = (\omega_o + \omega)(\omega_o - \omega) \\ \approx 2\omega_o(\omega_o - \omega). \tag{1.16}$$

Further noting that the quality factor of the RLC network is given by[1]

$$Q = \frac{\omega L}{R} \tag{1.17}$$

and making use of the identity of trigonometric functions

$$\tan^{-1}\left(x^{-1}\right) = \frac{\pi}{2} - \tan^{-1} x, \tag{1.18}$$

we arrive at

$$\tan\phi \approx -\frac{2Q}{\omega_o}(\omega_o - \omega). \tag{1.19}$$

For weak injection, ϕ is small,

$$\tan\phi \approx \phi \tag{1.20}$$

holds. Equation (1.19) can be approximated as

$$\phi \approx -\frac{2Q}{\omega_o}(\omega_o - \omega). \tag{1.21}$$

It follows that

$$A = \frac{d\phi}{d\omega} = \frac{2Q}{\omega_o}. \tag{1.22}$$

Equation (1.10) becomes

$$\frac{d\theta}{dt} + \Delta\omega_L = \frac{\omega_o}{2Q}\frac{I_{inj}}{I_o}\sin\theta. \tag{1.23}$$

When the frequency of the oscillator is the same as that of the injection signal, i.e., the oscillator is locked to the injection signal, their phase difference will be stationary. As a result,

[1]The parallel RLC network in Fig. 1.2 is derived from the network consisting of a series RL network depicting the spiral inductor and a parallel capacitor. The quality factor is dictated by the RL network.

$$\frac{d\theta}{dt} = 0. \tag{1.24}$$

Equation (1.23) in this case becomes

$$\Delta\omega_L = \frac{\omega_o}{2Q}\frac{I_{inj}}{I_o}\sin\theta. \tag{1.25}$$

$\Delta\omega_L$ is maximized when

$$\theta = \frac{\pi}{2}. \tag{1.26}$$

The maximum lock range is given by

$$\Delta\omega_{L,max} = \frac{\omega_o}{2Q}\frac{I_{inj}}{I_o}. \tag{1.27}$$

$\Delta\omega_{L,max}$ is termed the maximum lock range or simply the lock range of the oscillator. Clearly the lock range of the oscillator is inversely proportional to the quality factor of the oscillator and directly proportional to the relative injection strength I_{inj}/I_o of the oscillator.

Adler's theoretical treatment of the injection-locking of oscillators was validated experimentally by Huntoon and Weiss [31]. The theoretical results obtained by Adler assumed that the injection signal was a continuous wave. Fraser extended Adler's theory on the injection-locking of oscillators to oscillators under the injection of a periodically interrupted wave [21]. Mackey demonstrated experimentally that Adler's theory of the injection-locking of oscillators also accurately describes the locking phenomena of microwave X-band reflex klystron oscillators [49].

1.5 Paciorek: General First-Harmonic Injection

Alder's formula on the lock range of injection-locked harmonic oscillators is valid only if the strength of the injection signal is significantly smaller as compared with the output of the oscillator, i.e., weak injection. The other two constraints upon which Adler's formula was derived are usually not of a concern. This is because the frequency of the external injection signal is always chosen to be within the lock range of the oscillator under injection for the given injection strength. For applications where the strength of the injection signal is not small, is Adler's formula of lock range still valid? To answer this question, in this section we briefly present Paciorek's formula on the lock range of injection-locked oscillators that removes the weak injection constraint imposed on Adler's formula. Paciorek's formula was developed nearly 20 years after the inception of Adler's formula [61].

We use the same LC oscillator shown in Fig. 1.2 to present Paciorek's general formula of the lock range of injection-locked oscillators. The law of sine gives

$$\frac{I_o}{\sin(\theta-\phi)}=\frac{I_{inj}}{\sin\phi}, \tag{1.28}$$

from which we obtain

$$\tan\phi=\frac{I_{inj}\sin\theta}{I_o+I_{inj}\cos\theta}. \tag{1.29}$$

Equation (1.15) gives the phase of the RLC network and is repeated here for convenience

$$\tan\phi=-\frac{R}{\omega L}\frac{\omega_o^2-\omega^2}{\omega_o^2}. \tag{1.30}$$

Since ω is in the vicinity of ω_o,

$$\omega_o^2-\omega^2\approx 2\omega_o(\omega_o-\omega). \tag{1.31}$$

Equation (1.30) can be simplified to

$$\tan\phi\approx\frac{2Q(\omega-\omega_o)}{\omega_o}. \tag{1.32}$$

Equating (1.29) and (1.32) and noting

$$\omega-\omega_o=(\omega-\omega_{inj})+(\omega_{inj}-\omega_o)=\Delta\omega+\Delta\omega_L \tag{1.33}$$

yield

$$\frac{d\theta}{dt}+\Delta\omega_L=\frac{\omega_o}{2Q}\frac{I_{inj}}{I_o}\frac{\sin\theta}{1+\frac{I_{inj}}{I_o}\cos\theta}. \tag{1.34}$$

Let us comment on the preceding results prior to further development:

(a) It is seen through the steps of the derivation of (1.34) that no any constraint was imposed. Equation (1.34) is therefore valid for both weak injection and strong injection. Note that in the derivation of Adler's formula, $I_{inj}\ll I_o$ was imposed.
(b) Equation (1.34) is similar to (1.23) except the denominator and can be simplified to (1.23) if $\frac{I_{inj}}{I_o}\cos\theta\ll 1$, i.e., weak injection.
(c) When the oscillator is locked to the injection signal, $\frac{d\theta}{dt}=0$ will hold. We therefore arrive from (1.34)

$$\Delta\omega_L = \frac{\omega_o}{2Q}\frac{I_{inj}}{I_o}\frac{\sin\theta}{1+\frac{I_{inj}}{I_o}\cos\theta}. \tag{1.35}$$

(d) The optimal θ^* at which the lock range is maximized can be obtained by letting

$$\frac{d\left(\Delta\omega_L\right)}{d\theta} = 0. \tag{1.36}$$

The result is given by

$$\cos\theta^* = -\frac{I_{inj}}{I_o}. \tag{1.37}$$

Substituting (1.37) into (1.35) yields the maximum lock range

$$\Delta\omega_{L,max} = \frac{\omega_o}{2Q}\frac{I_{inj}}{I_o}\frac{1}{\sqrt{1-\left(\frac{I_{inj}}{I_o}\right)^2}}. \tag{1.38}$$

If $I_{inj} \ll I_o$, Eq. (1.38) is simplified to

$$\Delta\omega_{L,max} \approx \frac{\omega_o}{2Q}\frac{I_{inj}}{I_o}. \tag{1.39}$$

Equations (1.39) and (1.39) show that Paciorek's maximum lock range formula given in (1.38) is simplified to Adler's maximum lock range formula given in (1.27) in the case of weak injection.

Paciorek's formula of the lock range of harmonic oscillators was also derived by Mirzaei et al. with a geometrical interpretation so as to provide a graphical interpretation of the locking process of harmonic oscillators [54, 56].

1.6 Miller: Frequency Regenerative Injection

In 1939, R. Miller proposed a frequency regenerative injection-locking scheme that can be used for injection-locked frequency multiplication/division [53]. Figure 1.3 shows the block diagram of an injection-locked oscillator that uses Miller's frequency regenerative injection-locking scheme. It consists of two nonlinear blocks f_1 and f_2 that are used to generate various frequency components, a multiplier that multiplies the output of the nonlinear block f_1 and that of the nonlinear block f_2, and a linear frequency discriminator that functions as a high-Q band-pass filter. The nonlinear blocks can be characterized with their Taylor series expansion

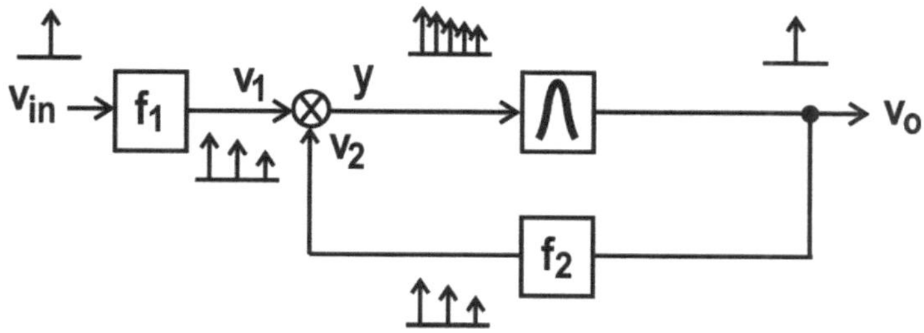

Fig. 1.3 Block diagram of an injection-locked oscillator using Miller's frequency regenerative injection-locking scheme

$$v_1 = f_1(v_{in}) \approx \sum_{m=1} a_m v_{in}^m, \tag{1.40a}$$

$$v_2 = f_2(v_o) \approx \sum_{n=1} b_n v_o^n. \tag{1.40b}$$

Let

$$v_{in} = V_{in} \cos(\omega_{in} t + \phi), \tag{1.41a}$$

and

$$v_o = V_o \cos(\omega_o t), \tag{1.41b}$$

we have

$$y = \left[\sum_{m=1} a_m V_{in}^m \cos^m(\omega_{in} t + \phi) \right] \left[\sum_{n=1} b_n V_o^n \cos^n(\omega_o t) \right]. \tag{1.42}$$

The mixer will generate various frequency components. Only those that are at ω_o will pass through the band-pass filter and reach the output of the oscillator. The oscillator can therefore perform either frequency multiplication if ω_o is a multiple of ω_{in} or frequency division if ω_o is a fraction of ω_{in}, depending upon the configuration of the nonlinear blocks. A further development of Miller's approach was given by Verma et al. where the two nonlinear blocks in Fig. 1.3 were combined into one so as to account for any interaction between them [83].

1.7 Schmideg: Superharmonic and Subharmonic Injection

The approaches of Adler and Paciorek for analysis of injection-locked harmonic oscillators are similar except that Paciorek's approach removes the weak inversion constraint imposed on Adler's formula. As a result, Paciorek's formula is valid for both weak and strong injection. Both Adler and Paciorek's approaches can only be used to analyze the first-harmonic injection of harmonic oscillators, i.e., the frequency of the injection signal needs to be located close to the natural frequency of the oscillator under injection. They cannot be used to analyze both superharmonic injection where the frequency of the injection signal is a superharmonic of the natural frequency of the oscillator or subharmonic injection where the frequency of the injection signal is a subharmonic of the natural frequency of the oscillator. In 1971, Schmideg proposed a first-harmonic balance method capable of analyzing injection-locked harmonic oscillators under the injection of either a superharmonic or a subharmonic signal [68]. Schmideg's approach forms the basis for analysis of injection-locked frequency dividers [63, 83]. In this section, we briefly present Schmideg's approach.

As mentioned earlier that a harmonic oscillator can be partitioned into a nonlinear block that performs frequency mixing and a high-Q band-pass filter that performs frequency discrimination, as shown in Fig. 1.4. Let the output of the oscillator in the lock state be a single tone of frequency ω in the vicinity of the passband center frequency ω_o of the bandpass filter, i.e.,

$$v_o = V_o \cos(\omega t) = \frac{V_o}{2}\left(e^{j\omega t} + e^{-j\omega t}\right). \tag{1.43}$$

Further let the injection signal be a single tone whose frequency is the kth harmonic of the output of the oscillator and the phase of the injection signal be ϕ.

$$v_{inj} = V_{inj}\cos(k\omega t + \phi) = \frac{V_{inj}}{2}\left(e^{jk\omega t}e^{j\phi} + e^{-jk\omega t}e^{-j\phi}\right). \tag{1.44}$$

The nonlinear block is depicted by the Taylor series expansion of the governing equation of the nonlinearity

$$y = a_1 x + a_2 x^2 + a_3 x^3 + \cdots . \tag{1.45}$$

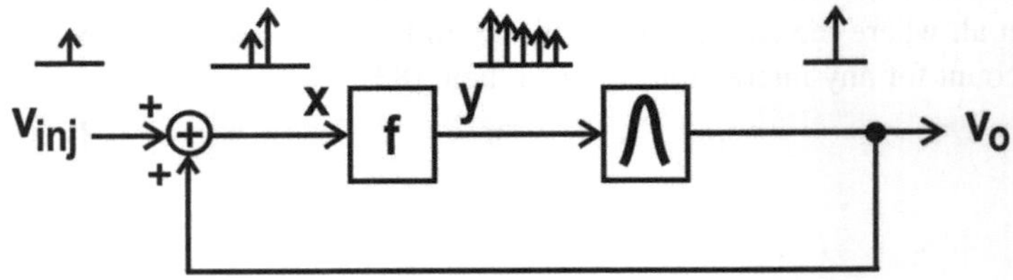

Fig. 1.4 Schmideg's model of injection-locked harmonic oscillators

The output of the nonlinear block of the oscillator is obtained by substituting (1.43) and (1.44) into (1.45).

$$\begin{aligned} y = a_1 &\left[\frac{V_{inj}}{2}\left(e^{jk\omega t}e^{j\phi} + e^{-jk\omega t}e^{-j\phi}\right) + \frac{V_o}{2}\left(e^{j\omega t} + e^{-j\omega t}\right)\right] \\ &+ a_2\left[\frac{V_{inj}}{2}\left(e^{jk\omega t}e^{j\phi} + e^{-jk\omega t}e^{-j\phi}\right) + \frac{V_o}{2}\left(e^{j\omega t} + e^{-j\omega t}\right)\right]^2 \\ &+ a_3\left[\frac{V_{inj}}{2}\left(e^{jk\omega t}e^{j\phi} + e^{-jk\omega t}e^{-j\phi}\right) + \frac{V_o}{2}\left(e^{j\omega t} + e^{-j\omega t}\right)\right]^3 \\ &+ \ldots . \end{aligned} \tag{1.46}$$

Among the frequency components in (1.46), only those that fall into the passband of the band-pass filter pass through the filter and reach the output of the oscillator while all other frequency components are suppressed by the band pass filter. The transfer function of the band-pass filter is given by

$$H_{BP}(j\omega) = \frac{H_o}{1 + j2Q\left(\frac{\omega-\omega_o}{\omega_o}\right)}, \tag{1.47}$$

where H_o is the gain of the filter at the passband center frequency ω_o and Q is the quality factor of the filter.

For the sake of simplicity, let us consider the simplest case where $k = 1$ (first-harmonic injection) and $y = a_1 x$. Focusing on frequency components associated with $e^{j\omega t}$, we have the following phasor relation

$$\frac{V_o}{2} \approx -\frac{H_o}{1 + j2Q\left(\frac{\omega-\omega_o}{\omega_o}\right)}\frac{a_1}{2}\left(V_{inj}e^{j\phi_{inj}} + V_o\right). \tag{1.48}$$

Separating the real and imaginary parts of (1.48) yields

$$1 + a_1 H_o + a_1 H_o \frac{V_{inj}}{V_o}\cos\phi_{inj} = 0, \tag{1.49}$$

and

$$\Delta\omega_L = \omega - \omega_o = \frac{\omega_o}{2Q}\frac{V_{inj}}{V_o}(a_1 H_o)\sin\phi_{inj}. \tag{1.50}$$

Equation (1.50) shows that $\Delta\omega_L$ will be maximized when

$$\phi^*_{inj} = \frac{\pi}{2}. \tag{1.51}$$

The maximum lock range is given by

$$\Delta\omega_{L,max} = \frac{\omega_o}{2Q}\frac{V_{inj}}{V_o}(a_1 H_o). \tag{1.52}$$

At $\phi^*_{inj} = \frac{\pi}{2}$, Eq. (1.49) becomes

$$1 + a_1 H_o = 0. \tag{1.53}$$

Equation (1.53) is Barkhausen criteria. Substituting (1.53) into (1.52) yields

$$\Delta\omega_{L,max} = -\frac{\omega_o}{2Q}\frac{V_{inj}}{V_o}. \tag{1.54}$$

It is seen that (1.54) is similar to (1.39) except in (1.39) the injection signal and the output of the oscillator are currents whereas in (1.54) the injection signal and the output of the oscillator are voltages.

It is evident that Schmideg's approach was derived using an approach that is more general and completely different from the approaches of Adler and Paciorek. It is capable of handling both a single injection signal and multiple injection signals. In addition, it is capable of handling both unitone injections and multitone injections as (1.44) can be readily modified to accommodate them [57]. Moreover, it is capable of handling superharmonic injection and subharmonic injection [110]. Furthermore, it is capable of handling an arbitrary degree of nonlinearity by properly truncating the Taylor series expansion of the nonlinear characteristic equation of the nonlinearity.

1.8 Progressive Multiphase Injection

For most harmonic oscillators, the number of injection signals is typically limited to two as these oscillators normally use a pair of cross-coupled common-source tuned amplifiers to generate required $-360°$ phase shift and synthesize a negative resistor, which is needed to compensate for the loss of the resonator so as to sustain oscillation. Unlike harmonic oscillators, a ring oscillator typically has more than two stages with the number of stages often set by the number of phases needed for a particular application. The presence of the multiple delay stages in ring oscillators makes us wonder whether only one external signal should be injected into a particular node of the oscillator or the number of external injection signals should be the same as that of the delay stages of the oscillator so as to maximize the lock range. If multiple injections indeed outperform single injection in terms of lock range, is there an optimal injection sequence or equivalently an optimal injection phase of each injection signal that yields the maximum lock range? To answer these questions, in this section we follow the approach of A. Mirzaei et al. [55].

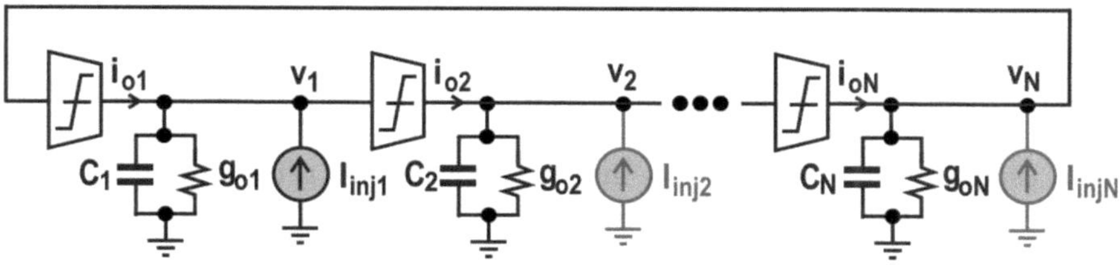

Fig. 1.5 Progressive multiphase injection in ring oscillators

Consider the ring oscillator shown in Fig. 1.5. The load capacitor of each stage is made of the output capacitance of the driving stage, the input capacitance of the driven stage, and the capacitance of the interconnect connecting the driving and driven stages. The load resistance comes from the output resistance of the driving stage only as the input resistance of the driven stage is infinite. Let us first assume that there is only one current i_{inj1} injected into node 1 of the oscillator and all other nodes have no injection. Assume that the output currents of all transconductors are the same. Let the per-stage delay of the oscillator be τ without injection. Note that τ is the average propagation delay of the delay stage. The free-running period of the oscillator is given by $T_o = N\tau$ where N is the number of the stages of the oscillator. Let the injection signal be a square wave of period T_{inj}. Since C_1 is charged by both i_{o1} and i_{inj1}, v_1 will rise faster as compared with the voltage of other nodes. As a result, the per-stage delay of stage 1 will be $\tau + \Delta\tau_1$ where $\Delta\tau_1 < 0$ is the variation of the delay of stage 1 caused by i_{inj1}. For the sake of simplicity, we assume that the injection current impacts the rising and falling edges equivalently.[2] Since in the lock stage, the period of the oscillator will be the same as that of the injection signal, we have

$$(N-1)\tau + (\tau + \Delta\tau_1) = T_{inj} \tag{1.55}$$

or

$$\Delta\tau_1 = T_{inj} - T_o. \tag{1.56}$$

The lock range is therefore given by

$$\begin{aligned}\Delta\omega &= \omega_{inj} - \omega_o \\ &= \frac{2\pi\left(T_o - T_{inj}\right)}{T_o T_{inj}} \\ &= -\left(\frac{\Delta\tau_1}{T_{inj}}\right)\omega_o.\end{aligned} \tag{1.57}$$

[2]Such an assumption is clearly not always correct since during the locking process, the period of the oscillator differs from that of the injection signal.

It becomes evident in (1.57) that the lock range is directly proportional to injection-induced delay variation of stage 1, which is proportional to i_{inj1}, specifically the duration in which C_1 is charged by i_{inj1}. Clearly one can increase the lock range by injecting i_{inj2} into node 2 when C_2 is charged by i_{o2}, i_{inj3} into node 3 when C_3 is charged by i_{o3}, ..., and i_{injN} into node N when C_N is charged by i_{oN}. This is known as progressive multiphase injection as the injection time of kth injection current $i_{inj,k}$ needs to be properly set such that the current injected into each node speeds up the charging process of the load capacitor of the node.

If each stage has one injection current and the injection phase of the injection current is set in a preceding progressive fashion, the load capacitor of each stage will be charged by both the output current of the transconductor and the injection current. We continue our early assumption that the injection current of each delay stage of the oscillator impacts the rising and falling edges of the output voltage of the stage equivalently. Since in the lock stage the period of the oscillator will be the same as that of the injection signal, we have

$$(\tau + \Delta\tau_1) + (\tau + \Delta\tau_2) + \cdots + (\tau + \Delta\tau_N) = T_{inj}. \tag{1.58}$$

As a result,

$$T_{inj} - T_o = \Delta\tau_1 + \Delta\tau_2 + \cdots + \Delta\tau_N. \tag{1.59}$$

The lock range is therefore given by

$$\begin{aligned} \Delta\omega &= \omega_{inj} - \omega_o \\ &= \frac{2\pi\left(T_o - T_{inj}\right)}{T_o T_{inj}} \\ &= -\left(\frac{\Delta\tau_1 + \Delta\tau_2 + \cdots + \Delta\tau_N}{T_{inj}}\right)\omega_o. \end{aligned} \tag{1.60}$$

Equation (1.60) shows that N-stage progressive multiphase injection increases the lock range by N times if the change of the per-stage delay induced by the injection signals is identical. Since for a given ring oscillator the number of the delay stages of the oscillator is known, one can properly assign the injection phase of each injection signal according to the delay of each stage of the oscillator so that each injection signal is progressively injected into the oscillator to achieve the maximum lock range.

The ring oscillator with progressive multiphase injection can be analyzed by assuming that the injection current at kth node is given by

$$i_{inj,k}(t) = I_{inj} e^{j(\omega_{inj} t + \phi_{inj,k})}, \tag{1.61}$$

where the injection phase is given by

$$\phi_{inj,k} = \omega_{inj}t - \frac{k-1}{N}\pi. \tag{1.62}$$

If the injection current is a square wave, only the first harmonic of the square wave whose frequency is ω_{inj} is considered as all other harmonics are assumed to be sufficiently suppressed by the RC network at the output of the delay stages of the oscillator. There are two currents flowing into the output node of each transconductor representing the delay stage: One is the output current of the transconductor and the other is the injection current. Since both the amplitude and phase of the voltage of the node vary with time, the output voltage of the kth transconductor can be written as

$$v_k = A(t)e^{j\phi(t)}, \tag{1.63}$$

where $A(t)$ and $\phi(t)$ are the amplitude and phase of the voltage of the node, respectively. Writing KCL at the output node of each transconductor in the time domain yields a total of N differential equations of the phase of the voltage of each node of the oscillator. They are termed generalized Adler's equations of ring oscillators. No closed-form solutions, however, were given in [55].

In [75, 76], Tofangdarzade et al. further developed the progressive multiphase injection approach developed by Mirzaei et al. and derived the closed-form solution of the generalized Adler's equations of ring oscillators and multiphase LC oscillators in the lock state and validated the obtained closed-form solutions using simulation results. Readers are referred to the cited references for the details of this approach.

Progressive multiphase injection was also developed for LC oscillators that can be treated as two-stage ring oscillators with each stage composed of a nonlinear transconductor and a RLC network. Using a negative feedback system approach, Yuan and Zhou showed that the lock range of the oscillator with a single injection signal is smaller as compared with that with dual injection signals. Also the lock range of the oscillator with dual injection signal depends upon the injection phases of the injection signals. The lock range is maximized once the difference between the injection phase of the injection signals is 180° [105, 112].

Mirzaei's progressive multiphase injection was originally developed for first-harmonic injection, i.e., the frequency of the progressively phased injection signals is in the vicinity of the frequency of the first harmonic of ring oscillators under injection. A. Musa et al. further developed the progressive mixing technique to injection-locked ring oscillator frequency dividers so as to obtain a large lock range [58] . Readers are referred to the cited references for the details of this approach.

1.9 Effective Injection Signaling

It has been well recognized that the nonlinear characteristics of oscillators, both harmonic oscillators and nonharmonic oscillators, are the root cause of injection-locking phenomena in oscillators. It has also been widely observed that the severer the degree of the nonlinearity of the oscillator under injection, the larger the lock range of the oscillator. In order to seek the theoretical explanation of this fundamental characteristic of oscillators, Yuan and Zhou utilized a Volterra series approach to quantify the impact of the degree of the nonlinearity of oscillators on their lock range [106, 107, 113]. In this approach, the nonlinearity of the oscillator under injection is depicted by its truncated Taylor series expansion first. The variables of the oscillator such as nodal voltages and branch currents are then represented by their truncated Volterra series expansion, thereby converting the nonlinear differential equations depicting the oscillator to a set of linear differential equations characterizing a set of linear circuits called Volterra circuits. Each of the Volterra circuits represents a harmonic oscillator and the frequencies of these harmonic oscillators are harmonically related. The input of high-order Volterra circuits is obtained from the response of lower-order Volterra circuits. The response of the oscillator is obtained by summing up that of the Volterra circuits. By solving Volterra circuits sequentially starting from the first-order Volterra circuit, the response of the oscillator can be obtained. Since the frequency of the first-order Volterra circuit is the same as that of the first harmonic of the oscillator, one only needs to find the lock range of the first-order Volterra circuit in order to obtain the lock range of the oscillator. The total or effective injection signals of the first-order Volterra circuit consist of two distinct parts: One from the external injection signals and the other contributed by higher-order Volterra circuits. It is the contribution of higher-order Volterra circuits that strengthens the effective injection signal of the first-order Volterra circuit subsequently a larger lock range. The severer the degree of the nonlinearity of the oscillator, the more the higher-order Volterra circuits contribute to the effective injection signal of the first-order Volterra circuit, the larger the lock range of the oscillator. The details of this approach will be provided in the later chapters.

1.10 Applications of Injection-Locking of Oscillators

The injection-locking of oscillators has found a broad range of applications in mixed-mode signal processing. In this section, we briefly browse through some of the representative applications of the injection-locking of oscillators. This, by no means, is an attempt to include all key applications of the injection-locking of oscillators. As the design of systems utilizing the injection-locking of oscillators is not the focus of this monograph, no details on the operation of these systems are provided. Readers are referred to the cited references for the detailed information of the principle and operation of these systems.

1.10.1 Frequency Division

Perhaps the most representative application of the injection-locking of oscillators is the pre-scalar of frequency synthesizers in both RF communication systems and digital systems. In an integer-N frequency synthesizer, a frequency divider with a large division ratio is needed in the feedback path to generate a low-frequency signal from the output of the high-frequency oscillator of the synthesizer so that the frequency of the low-frequency signal can be compared with that of a low-frequency reference, often from a crystal oscillator, such that the frequency of the oscillator can be stabilized. A large frequency division ratio is often obtained using a multistage approach in which multiple frequency dividers are cascaded. The first stage is most difficult to design and also consumes the most power. This is because not only the first-stage must operate at the same frequency as that of the oscillator of the synthesizer, it must also cope with the large variation of the frequency of its input arising from the impact of PVT (process, voltage, and temperature) uncertainty on the frequency of the oscillator of the synthesizer. It becomes increasingly difficult to design frequency dividers operating at high frequencies using conventional digital logic such as current-mode logic (CML) not because of their speed limitation but also due to their excessive dynamic power consumption. Attributive to the intrinsic characteristics of injection-locked oscillators including high frequency accuracy, fast locking process, low power consumption, and low phase noise once locked, injection-locked frequency dividers are attractive in realization of the first stage of the divide-by-N block of frequency synthesizers known as the pre-scalar. Both injection-locked harmonic oscillator frequency dividers [8, 11, 14, 19, 33, 34, 74, 94, 99] and injection-locked nonharmonic oscillator frequency dividers are widely popular [18, 30, 47, 48, 98].

1.10.2 Parallel Links

High-bandwidth short-distance parallel data links are critical for multi-core processing and networking applications. Forwarded-clock (FC) also known as source synchronous architecture where the sampling clock of the receiver of a parallel data link is transmitted directly from the transmitter and multiple data channels share the same clock channel so as to minimize routing cost is widely favored due to the inherent jitter correlation between data and clock as both clock and data are generated by the same transmitter and the fact that the performance of data links is dominated by the relative jitter between clock and data rather than the absolute jitter of either clock or data [41]. Jitter correlation between clock and data is only beneficial over a certain frequency range called jitter transfer bandwidth beyond which tracking high-frequency jitter is detrimental.

As lossy channels amplify data jitter at high frequencies [5], filtering high-frequency data jitter beyond the jitter transfer bandwidth in which jitter correlation between data and clock exists is important. As delay mismatch between data and clock channels can be several unit intervals (UIs), which reduces jitter correlation between clock and data [70], minimizing channel mismatch via deskewing is critical. Channel mismatch induced clock and data skews also mandate that deskewing be performed at the receiver such that data can be sampled at the center of data eyes.

To lower power consumption and minimize interference from the clock channel, it is desirable to transmit a low-frequency clock over the clock channel and generate a high-frequency clock from the transmitted low-frequency clock at the receiver using a frequency multiplier. Frequency multiplication can be performed using a phase-locked loop (PLL) or a delay-locked loop (DLL). PLL frequency multiplication removes high-frequency data jitter, accredited to its finite loop bandwidth. It, however, also removes desired correlated jitter between data and clock that resides in jitter transfer bandwidth [9, 27]. The loop bandwidth, on the other hand, must be sufficiently large in order to allow PLLs to track data. This also helps retain the desirable correlated jitter. The loop bandwidth of PLLs, however, is limited to less than 1/10 of the frequency of the input due to loop stability constraints [46, 59]. On top of that, PLL-frequency multiplication suffers from phase accumulation. As a result, PLL frequency multiplication is not particularly attractive at high data rates. DLL frequency multiplication uses a multiplying delay-locked loop (MDLL) to generate a high-frequency clock from a low-frequency clock [40]. It is generally preferred over their PLL counterparts as phase accumulation only exists within one period of the input. This differs from PLL frequency multiplication where phase accumulation takes place over multiple periods of the input until it becomes sufficiently large such that it can be detected by the phase detector and subsequently corrected by the feedback loop. DLL frequency multiplication, however, suffers from jitter amplification at high frequencies, which undesirably increases high-frequency data jitter beyond jitter transfer bandwidth.

Zhang et al. proposed an injection-locking clock distribution scheme in which the frequency of local oscillators responsible for local clocking is synchronized to the global clock via injection-locking [108]. With superharmonic injection, the frequency of the global clock can be lowered, thereby lowering both power consumption and interference from global clock trees. As the injection signal need not to be large, the number of buffers needed to relay global clocks can also be lowered. This is echoed with both power consumption reduction and clock jitter improvement.

O'Mahony et al. proposed a forwarded-clock parallel I/O architecture with an injection-locked oscillator to generate required deskewing clocks at the receiver [60, 70]. The injection-locking signal is the clock transmitted by the clock channel. The injection-locked oscillator behaves as a low-pass filter that filters out unwanted high-frequency jitter beyond the desired jitter transfer bandwidth and duty-cycle errors present in the received clock. Since the inception of this source synchronous parallel data link architecture, numerous studies have been reported and the effectiveness of this architecture has been validated by measurement results [27–29].

1.10.3 Frequency Calibration

The system clock of a PWM such as an embedded temperature sensor or a radio-frequency identification (RFID) tag that controls the operation of both the RF and baseband blocks of the PWM can be sent directly by its base station either via its carrier or the envelope of the carrier when amplitude shift key (ASK) modulation scheme is used [36]. This approach suffers from a number of drawbacks: (a) High power consumption as the comparator in the clock recovery block must operate at the carrier frequency, which is significantly higher as compared with the frequency of the system clock of the PWM typically. The need for a chain of frequency dividers to generate the system clock whose frequency is typically much lower than that of the carrier also greatly increases power consumption. (b) A large modulation index is preferred from a data and clock recovery point of view. Because the amplitude of the received RF signal is usually small, especially in logic-0 states when the modulation index of the ASK-modulated RF signal is large, a large timing error will exist in the recovered clock. (c) The availability of the system clock in down-link operation only greatly restrains the flexibility of the system.

In order to avoid the drawbacks of generating the system clock of a PMW directly from its carrier, it is desirable to generate the system clock of the PWM locally using either a ring oscillator or a relaxation oscillator. The frequency of the local oscillator, however, is subject to the impact of PVT uncertainty and needs to be calibrated with reference to a clock of a known frequency prior to the intended operations of the PWM. The reference clock with respect to which the frequency of the PWM is to be calibrated is sent from the base station rather than generated locally using a timing oscillator whose frequency is independent of the effect of PVT uncertainty. This is because it is difficult to generate a precision band-gap reference voltage or current in the presence of a fluctuating supply voltage and varying temperature with the constraint of a limited amount of power consumption allowed.

Frequency calibration can be performed using a frequency-locked loop. FLL-based frequency calibration generally suffers from high power consumption, a low frequency accuracy, and a long calibration time, making it less attractive for PWMs. Frequency calibration using injection-locking with the frequency reference against which the frequency of the local oscillator is calibrated as the injection signal sent from the base station is an effective way to lower power consumption, ameliorate low frequency accuracy, and shorten calibration time while ensuring a good frequency calibration accuracy [3, 4, 37, 73].

1.10.4 Phase-Locked Loops

Low phase noise phase-locked loops (PLLs) are important building blocks for a broad range of applications including data links and frequency synthesizers. The phase noise of PLLs is dominated by either the phase noise of its input or the phase noise of its oscillator, application-dependent. For example, for the transmitter PLL of a serial data link, since the input of the PLL in this case is often a crystal oscillator whose phase noise is low, the phase noise of the PLL is dominated by that of its oscillator whereas for the receiver PLL of the serial data link, the phase noise of the PLL is dictated by the jitter of its input, which is the data received at the far end of the channel, specifically, the equalized data. One can manipulate the phase noise of a PLL by adjusting the loop bandwidth of the PLL [43]. For example, for the transmitter PLL, the loop bandwidth should be maximized so as to minimize the contribution of the phase noise of the oscillator to the overall phase noise of the PLL whereas for the received PLL, the loop bandwidth should be minimized in order to minimize the contribution of the jitter of the received data to the overall phase noise of the PLL. The deployment of these effective design techniques, however, is confronted by other constraints such as stability and lock time imposed on the PLL. For example, for the received PLL, the loop bandwidth of the PLL must also be maximized so as to allow the PLL to lock to incoming data in typically less than half of the unit interval (UI) or symbol time even though minimizing loop bandwidth is preferred from a phase noise point of view.

The spectrum of the phase noise of a PLL typically consists of three distinct sections: A low frequency offset section in which the phase noise of the PLL follows the profile of that of the input but is larger, mainly dominated by flicker noise, a high frequency offset section near and beyond the loop bandwidth where the phase noise of the PLL is nearly the same as that of the oscillator, and an intermediate frequency offset section up to the loop bandwidth where the phase noise of the PLL is nearly constant [16, 17, 24, 39, 42, 62, 72, 90, 109]. Lowering the phase noise of the PLL in both the low frequency offset and intermediate frequency offset sections, i.e., within the loop bandwidth of the PLL, is critical.

As to be shown in later chapters that injection-locking is capable of lowering the phase noise of an oscillator once the oscillator is locked to a clean reference signal, i.e., its phase noise is significantly lower than that of the free-running oscillator that is to be injection-locked. This unique characteristic of injection-locking has attracted a lot of attention recently as an effective means to lower the phase noise of PLLs without sacrificing the loop bandwidth of the PLLs. For the PLL of frequency synthesizers, as a clean injection-signal is provided by a crystal oscillator while the frequency of the oscillator of the PLL for wireless and wireline communications is in the range of multi-GHz, the subharmonic injection of the oscillator of the PLLs is generally performed [43]. Injection-locking should always be accompanied with the PLL to ensure lock as the lock range of oscillators is much smaller as compared that of the PLL such that injection-locking alone cannot guarantees lock. The injection-locking signal is typically a train of narrow pulses generated from the reference signal of the PLL using a pulse generator. As to be shown in the later chapters that injection will be most effective if injection-locking signal is injected at the threshold-crossing of the output of the oscillator under injection. The effectiveness of injection-locking critically depends upon the time at which the injection-locking pulse is applied to the oscillator. Timing mismatch between the time instant at which the injection signal is injected into the oscillator and the time instant at which the output of the oscillator crosses the threshold gives rise to reference spurs [72].

1.11 Chapter Summary

A general classification of oscillators was provided. We showed that oscillators can be loosely classified into harmonic oscillators and nonharmonic oscillators. The brief history of the injection-locking of oscillators was presented with an emphasis on the analysis rather than design of injection-locked oscillators. The first-harmonic method for analysis of harmonic oscillators in weak inversion developed by Adler was studied in detail. It was followed with the presentation of the first-harmonic method for analysis of harmonic oscillators in both weak and strong developed by Paciorek. Frequency regenerative injection method developed by Miller and further enhanced by other researchers for the realization of frequency multiplication and frequency division was explored. Schmideg's first-harmonic balance method capable of analyzing harmonic oscillators in first-harmonic, superharmonic, or subharmonic injections was studied. The progressive multiphase injection of ring oscillators was examined. We showed that the lock range of a ring oscillator with properly phased N injections of equal strength is N times that of the oscillator with only one injection. The further development of progressive multiphase injection for multiphase harmonic oscillators and injection-locked ring oscillator frequency dividers was referenced. The effective injection signaling arising from the nonlinearity of oscillators under injection and obtained by analyzing the Volterra circuits of the oscillators was described. Finally, the representative applications of the injection-locking of oscillator were briefly explored.

Chapter 2
Injection-Locking of Harmonic Oscillators

2.1 Introduction

Harmonic oscillators are those whose output is sinusoidal. Typical examples of harmonic oscillators are LC oscillators with either a spiral inductor/transformer resonator or an active inductor/transformer resonator. As compared with their nonharmonic counterparts such as ring oscillators or relaxation oscillators, harmonic oscillators feature better phase noise performance, rooted to the higher quality factor of the LC resonator subsequently a better frequency selectivity. Harmonic oscillators are widely used in wireless communication systems where a stringent constraint on the phase noise of oscillators exists. Injection-locked harmonic oscillators are often used as the pre-scalar of a frequency synthesizer to perform frequency division at high frequencies with which conventional digital frequency dividers fail to cope. This chapter is concerned with the fundamentals of the injection-locking of harmonic oscillators with an emphasis on lock range. Techniques that boost the lock range of injection-locked harmonic oscillators will be dealt with in the next chapter.

The chapter is organized as follows: The fundamentals of the oscillation of harmonic oscillators are studied in Sect. 2.2. Such a study is warranted as it lays a necessary foundation for an in-depth investigation of the mechanisms that determine both the locking behavior and lock range of harmonic oscillators. It is followed with a close examination of the noise spectrum of harmonic oscillators in Sect. 2.3. The modeling of injection-locked harmonic oscillators using a nonlinear negative feedback system is investigated in Sect. 2.4. The first-harmonic injection-locking of harmonic oscillators is investigated in Sect. 2.5. Both nonlinear and linear approaches capable of deriving the lock range of injection-locked harmonic oscillators are studied. We show that the nonlinear approach provides the much needed information of the locking process of harmonic oscillators, while the linear approach is capable of yielding the lock range of harmonic oscillator without the prior knowledge of the locking process of the oscillator. We further show that the lock range of harmonic oscillators can be quantified by the injection-induced

F. Yuan, *Injection-Locking in Mixed-Mode Signal Processing*,
https://doi.org/10.1007/978-3-030-17364-7_2

variation of the impedance of the resonator of the oscillators. Such an approach is of great importance as it reveals a distinct characteristic between harmonic oscillators with a spiral inductor resonator and those with an active inductor resonator. Section 2.6 investigates the superharmonic injection-locking of harmonic oscillators. Both second-order and third-order superharmonic injection-locking of harmonic oscillators are studied. Divide-by-2 and divide-by-3 injection-locked frequency dividers are the representative applications of the superharmonic injection-locking of harmonic oscillators and are studied in detail. The subharmonic injection-locking of harmonic oscillators is investigated in Sect. 2.7. The intrinsic relations among the lock range of harmonic oscillators in first-harmonic, superharmonic, and subharmonic injection-locking are explored in Sect. 2.8. Section 2.9 investigates the phase noise of injection-locked harmonic oscillators. The chapter is summarized in Sect. 2.10.

2.2 Harmonic Oscillators

Oscillation is a phenomenon of feedback systems. The oscillation of harmonic oscillators is often studied using a linear negative feedback system approach. This is because the output of a harmonic oscillator is a sinusoid ideally, attributive to the high-quality factor of the resonator of these oscillators. In contrast, the output of nonharmonic oscillators such as a ring oscillator or a relaxation oscillator is a square wave ideally. When a harmonic oscillator is locked to an external sinusoidal injection signal, the oscillator will oscillate at the frequency of the injection signal rather than the natural frequency of the oscillator. The existence of only one frequency component in the spectrum of the output of an injection-locked harmonic oscillator in the lock state permits the use of linear system analysis techniques such as Laplace transform to depict the operation of the oscillator in the lock state, as to be detailed in this chapter. Since multiple frequency components coexist prior to the establishment of the lock state, linear system approaches cannot be used to analyze the locking process of harmonic oscillators.

The closed-loop transfer function of a linear negative feedback system, denoted by H_c, is given by

$$H_c = \frac{H_o F}{1 + H_o F}, \tag{2.1}$$

where H_o is the open-loop transfer function of the system, and F is the transfer function of the feedback path. The system will oscillate at ω_o without any external injection signal, occurring when

$$1 + H_o(j\omega_o)F(j\omega_o) = 0 \tag{2.2}$$

is satisfied. Equation (2.2) is known as Barkhausen criteria, in memory of Germany physicist Heinrich Georg Barkhausen (1881–1956). ω_o is known as the natural or free-running frequency of the system. Note that since $H_o(j\omega_o)$ is a complex variable typically, Barkhausen criteria consist of a magnitude criterion

$$|H_o(j\omega_o)F(j\omega_o)| = 1 \tag{2.3}$$

and a phase criterion

$$\angle\left[H_o(j\omega_o)F(j\omega_o)\right] = \pm 360^\circ. \tag{2.4}$$

The magnitude and phase criteria need to be satisfied simultaneously in order for the system to oscillate. Since electronic systems typically assert a negative phase shift and the negative feedback mechanism guarantees a -180° phase shift, Eq. (2.4) can be simplified to

$$\angle\left[H_o(j\omega_o)F(j\omega_o)\right] = -180^\circ. \tag{2.5}$$

Let us consider the tuned amplifier shown in Fig. 2.1. It is a common-source amplifier with a spiral inductor resonator load. The resonator is modeled as a RLC network where L_s denotes the inductance of the spiral inductor, R_s denotes the series resistance of the spiral, and C_p denotes the capacitance. Note that C_p consists of the capacitance of the tuning capacitor for adjusting the resonant frequency of the resonator, the capacitance of the MOS transistor seen at the drain of the transistor, and the parasitic capacitances of the spiral. As the capacitance of the transistor seen at its drain is absorbed by the RLC network while the capacitance at the gate of the transistor is absorbed by the input, the transistor functions as a transconductor with

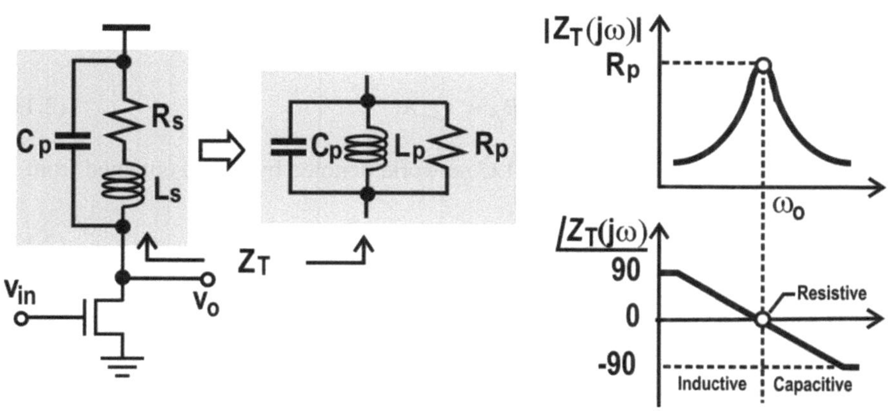

Fig. 2.1 Left: Common-source amplifier with a RLC resonator load. Right: The magnitude and phase of the impedance of RLC resonator

output conductance $g_o = 1/r_o$ where r_o is the output resistance of the transistor. To facilitate analysis, the series RL network can be replaced with the parallel RL network shown in the figure at a specific frequency, typically the resonant frequency of the LC resonator $\omega_o = 1/\sqrt{L_s C_p}$ under the condition that the impedance of the series RL network is the same as that of the parallel RL network, i.e.,

$$R_s + j\omega_o L_s = R_p||(j\omega_o L_p), \tag{2.6}$$

where operator || indicates the parallel connection of the components on both sides of the operator, and subscripts s and p specify series and parallel connections, respectively. Equating the real and imaginary parts of the expressions on both sides of (2.6) yields

$$L_p = L_s\left(1 + \frac{1}{Q_s^2}\right), \tag{2.7}$$

and

$$R_p = R_s(1 + Q_s^2), \tag{2.8}$$

where

$$Q_s = \frac{\omega L_s}{R_s} \tag{2.9}$$

is the quality factor of the spiral inductor. If Q_s is sufficiently large, which is typically the case, we will have

$$L_p \approx L_s \tag{2.10}$$

and

$$R_p \approx Q_s^2 R_s. \tag{2.11}$$

The impedance of the parallel RLC network, denoted by Z_T, is obtained from

$$Z_T = R_p||j\omega L_p||\frac{1}{j\omega C_p}. \tag{2.12}$$

The magnitude of Z_T, denoted by $|Z_T|$, and the phase of Z_T, denoted by $\angle Z_T$, are sketched in Fig. 2.1. It is seen that the RLC network will exhibit an inductive characteristic when $0 < \omega < \omega_o$, a resistive characteristic at $\omega = \omega_o$, and a capacitive characteristic when $\omega > \omega_o$.

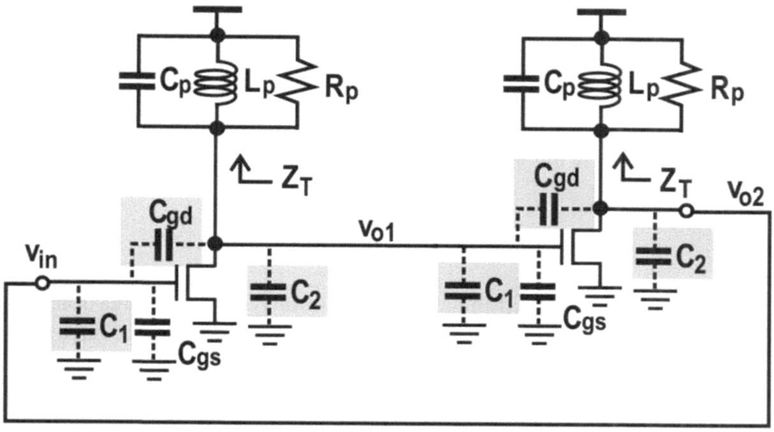

Fig. 2.2 Harmonic oscillators with LC resonators

At the resonant frequency of the RLC network, the network becomes purely resistive with resistance $Z_T = R_p||r_o$. The amplifier in this case is an inverting amplifier with voltage gain

$$A_v = -g_m(r_o||R_p), \tag{2.13}$$

where g_m and r_o are the transconductance and output resistance of the transistor, respectively. The amplifier provides a $-180°$ phase shift. The gain of the amplifier at ω is given by

$$A_v = -g_m \left[r_o||Z_T(\omega)\right]. \tag{2.14}$$

Because $|Z_T|\leq R_p$, the gain is maximized at ω_o. It will drop when $\omega \neq \omega_o$ and become zero when $\omega = 0$ or $\omega = \infty$.

Since the maximum phase shift of the amplifier is $-180°$, occurring at the self-resonant frequency of the LC resonator, in order to satisfy Barkhausen phase criterion, two identical tuned amplifiers can be cascaded in series with the output of the second amplifier routed back to the input of the first amplifier, as shown in Fig. 2.2. As each stage provides a $-180°$ phase shift at ω_o, a total of $-360°$ phase shift can be obtained at ω_o, ensuring that Barkhausen phase criterion is satisfied at ω_o.

To simplify analysis, we neglect the parasitic capacitances of the transistors and only concentrate on the intrinsic capacitances of the transistors, i.e., the gate-source capacitance C_{gs} and gate-drain capacitance C_{gd} of the transistors. Using Miller's theorem, floating capacitor C_{gd} is replaced with two single-ended capacitors [65]

$$C_1 = (1 - A_v)C_{gd} \approx -A_v C_{gd} \tag{2.15}$$

at the gate and

$$C_2 = \frac{C_{gd}}{1 - A_v^{-1}} \approx C_{gd} \tag{2.16}$$

at the drain where A_v is the voltage gain from the gate to the drain of the transistors and is given by

$$A_v \approx -g_m \left(r_o || Z_T\right). \tag{2.17}$$

Note that we have assumed that $A_v \gg 1$ in simplifying the expression of C_1 and that of C_2. It is evident that the total capacitance at the gate of each transistor, denoted by $\hat{C}_p$, is given by

$$\hat{C}_p = C_p + C_1 + C_2 + C_{gs}. \tag{2.18}$$

The gain of the cascaded amplifiers or the open-loop voltage transfer function of the negative feedback system depicting the oscillator is given by

$$H_o = A_v^2 = g_m^2 \left(r_o || \hat{Z}_p\right)^2, \tag{2.19}$$

where

$$\hat{Z}_p = R_p || j\omega L_p || 1/(j\omega \hat{C}_p). \tag{2.20}$$

Once the oscillator oscillates, it will only oscillate at the resonant frequency of the resonator ω_o. All other frequency components generated by the nonlinear characteristics of the transistors will vanish provided the quality factor of the resonator is infinite. The reason for this is two-fold: (a) At ω_o, since the phase shift of each stage is $-180°$, the total phase shift is $-360°$, thereby satisfying Barkhausen phase criterion and (b) at ω_o, the impedance of the resonator is maximized so is the voltage gain, ensuring that Barkhausen magnitude criterion is also satisfied.

At frequencies other than ω_o, since $|Z_T(j\omega)| \ll |Z_T(j\omega_o)|$ due to the high-quality factor of the resonator, the gain of each stage drops greatly. Also, the phase shift provided by each stage is now less than $-180°$. As a result, both Barkhausen magnitude and phase criteria will not be satisfied. No oscillation at these frequency will therefore take place.

Although satisfying Barkhausen criteria guarantees the oscillation of the oscillator, in order to sustain oscillation, the energy loss of the resonator must be compensated. The energy loss of the resonator arises mainly from (a) the resistance of the spirals and (b) the resistance of the varactors used for frequency tuning. The existence of the resistive loss of the resonator requires that additional energy be pumped into the system continuously to compensate for the loss of energy so as to sustain oscillation. This is achieved using a negative resistor realized using a pair of cross-coupled NMOS transistors, as shown in Fig. 2.3. Since both RLC networks share the same supply voltage, they can be merged into one RLC network with

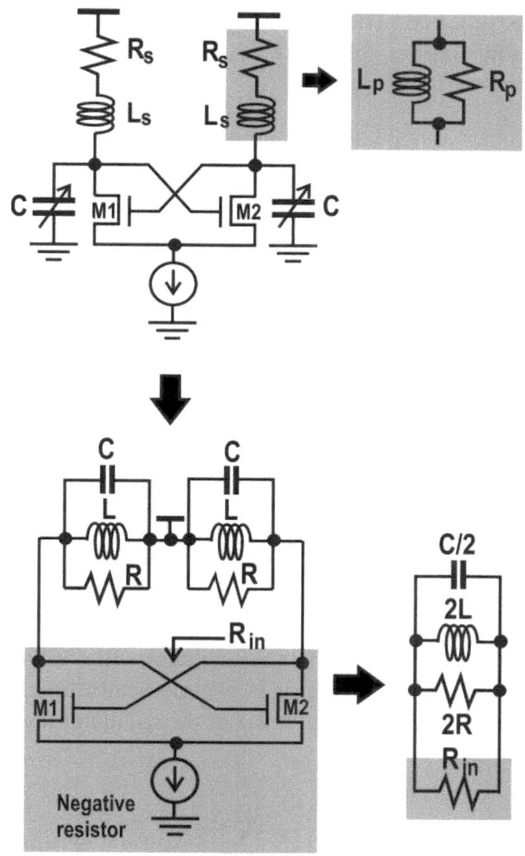

Fig. 2.3 Harmonic oscillator with spiral inductor resonators. The cross-coupled transistors M1 and M2 form a negative resistor that compensates for the resistive loss of the resonator

inductance $2L$, capacitance $C/2$, and resistance $2R$, as shown in Fig. 2.3. Note that the merged RLC network has the same resonant frequency as that of individual LC resonator. The impedance looking into the drains of the cross-coupled NMOS pair can be conveniently derived using a small-signal equivalent circuit approach without worrying about C_{gs}, C_{gd}, and r_o of the transistors. This is because C_{gs}, C_{gd}, and r_o have already been absorbed into the RLC networks, as shown earlier. It can be shown that the resistance looking into the drains of the cross-coupled transistors, denoted by R_{in}, is given by

$$R_{in} \approx -\frac{2}{g_m} \tag{2.21}$$

In order to sustain oscillation,

$$R_{in} + 2R = 0 \tag{2.22}$$

is required. The resistance of the synthesized negative resistor can be tuned by adjusting g_m obtained by varying the gate voltage of the trail current source transistor because

$$g_m \approx \mu_n C_{ox} \left(\frac{W}{L}\right) (V_{GS} - V_{Tn}). \tag{2.23}$$

where μ_n is the surface mobility of free electronics, C_{ox} is gate capacitance per unit area, W and L are the width and length of the transistor, respectively, V_{GS} is gate-source DC biasing voltage, and V_{Tn} is the threshold voltage of NMOS transistors.

2.3 Phase Noise of Oscillators

In this section, we investigate the noise spectrum of harmonic oscillators. Consider a harmonic oscillator with its free-running frequency ω_o. Let there be a noise current i_n whose power spectral density is given by $S_{in}(\omega)$. i_n is injected into the oscillator, as shown in Fig. 2.4. The injection of i_n causes the frequency of the oscillator to shift from its free-running frequency ω_o to $\omega_o + \Delta\omega$ with $\Delta\omega \ll \omega_o$ typically. Note that frequency shift $\Delta\omega$ will be needed in order to satisfy Barkhausen criteria when i_n is present such that the oscillator will oscillate at $\omega_o + \Delta\omega$.

Since the oscillator is a periodically time-varying system, the power spectral density of the output noise of the oscillator at $\omega_o + \Delta\omega$, denoted by $S_o(\omega_o + \Delta\omega)$, is given by [101]

$$S_o(\omega_o + \Delta\omega) = \sum_{k=-\infty}^{\infty} |H_c(\omega_o + \Delta\omega, k\omega_o + \Delta\omega)|^2 \, S_{in}(k\omega_o + \Delta\omega), \tag{2.24}$$

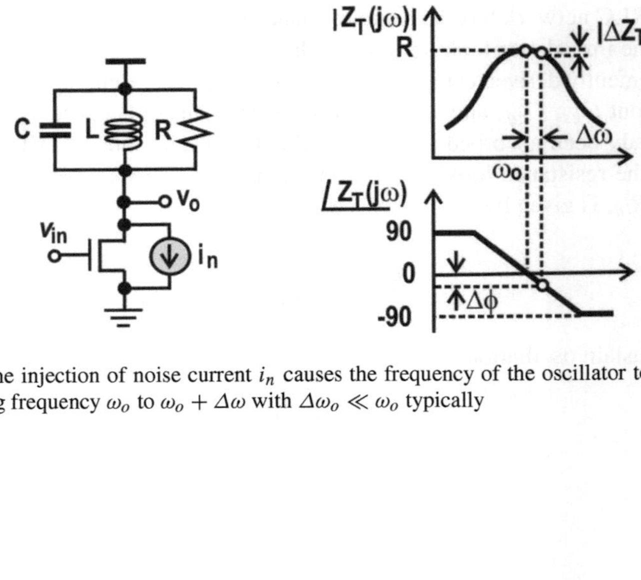

Fig. 2.4 The injection of noise current i_n causes the frequency of the oscillator to shift from its free-running frequency ω_o to $\omega_o + \Delta\omega$ with $\Delta\omega_o \ll \omega_o$ typically

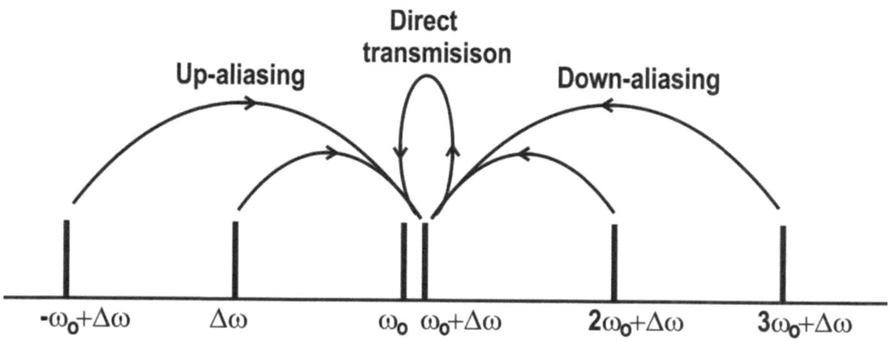

Fig. 2.5 Aliasing transfer function

where $H_c(\omega_o + \Delta\omega, k\omega_o + \Delta\omega)$ is the aliasing transfer function of the oscillator with input at frequency $k\omega_o + \Delta\omega$ and output at frequency $\omega_o + \Delta\omega$. Note that the frequency of the input and that of the output might differ. To better understand the aliasing transfer functions, we rewrite (2.24) in the following form

$$\begin{aligned} S_o(\omega_o + \Delta\omega) = \ldots &+ |H_c(\omega_o + \Delta\omega, \Delta\omega)|^2 S_{in}(\Delta\omega) \\ &+ |H_c(\omega_o + \Delta\omega, \omega_o + \Delta\omega)|^2 S_{in}(\omega_o + \Delta\omega) \\ &+ |H_c(\omega_o + \Delta\omega, 2\omega_o + \Delta\omega)|^2 S_{in}(2\omega_o + \Delta\omega) + \cdots \end{aligned} \tag{2.25}$$

It is seen from (2.25) that the first term in (2.25) quantifies the up-aliasing of the noise from $\Delta\omega$ to $\omega_o + \Delta\omega$, the second term depicts the direct transmission of the noise at $\omega_o + \Delta\omega$, and the third-term describes the down-aliasing of the noise from $2\omega_o + \Delta\omega$ to $\omega_o + \Delta\omega$. Figure 2.5 provides a graphical presentation of the aliasing transfer function.

Since the PSD of the thermal noise of the channel current of MOS transistors, the PSD of the flicker noise of the channel current of MOS transistors, and the PSD of the thermal noise of resistors are known, one only needs to know the aliasing transfer function $H_c(\omega_o + \Delta\omega, k\omega_o + \Delta\omega)$ in order to obtain the power spectral density of the output of the oscillator.

Since $\Delta\omega \ll \omega_o$, the following first-order approximation holds

$$H_o(\omega_o + \Delta\omega) \approx H_o(\omega_o) + \left(\frac{dH_o}{d\omega}\right)_{\omega_o} \Delta\omega. \tag{2.26}$$

Because

$$H_o(j\omega_o) = -g_m Z_T(j\omega_o) = -g_m R, \tag{2.27}$$

and

$$\left[\frac{dZ_T(j\omega)}{d\omega}\right]_{\omega_o} = 0 \tag{2.28}$$

typically, as well as at $\omega_o + \Delta\omega$,

$$H_o(\omega_o + \Delta\omega) \approx H_o(\omega_o) = -1 \tag{2.29}$$

holds, as illustrated graphically in Fig. 2.4, we have

$$\begin{aligned} H_c(\omega_o + \Delta\omega) &= \frac{H_o(\omega_o + \Delta\omega)}{1 + H_o(\omega_o + \Delta\omega)} \\ &\approx \frac{-1}{\left(\frac{dH_o}{d\omega}\right)_{\omega_o}(\Delta\omega)}. \end{aligned} \tag{2.30}$$

It follows that

$$|H_c(\omega_o + \Delta\omega)|^2 \approx \frac{1}{\left|\frac{dH_o}{d\omega}\right|^2_{\omega_o}(\Delta\omega)^2}, \tag{2.31}$$

where $|x|$ returns the magnitude of complex variable x. Let

$$H_o = Ae^{j\phi}, \tag{2.32}$$

where A and ϕ denote the magnitude and phase of H_o. We have

$$H_o^* = Ae^{-j\phi}. \tag{2.33}$$

It follows that

$$\frac{dH_o}{d\omega} = \frac{dA}{d\omega}e^{j\phi} + Ae^{j\phi}\frac{d\phi}{d\omega}, \tag{2.34a}$$

$$\frac{dH_o^*}{d\omega} = \frac{dA}{d\omega}e^{-j\phi} - Ae^{-j\phi}\frac{d\phi}{d\omega}. \tag{2.34b}$$

As a result,

$$\begin{aligned} \left|\frac{dH_o}{d\omega}\right|^2_{\omega_o} &= \left(\frac{dH_o}{d\omega}\frac{dH_o^*}{d\omega}\right)_{\omega_o} \\ &= \left(\frac{dA}{d\omega}\right)^2_{\omega_o} + A^2(\omega_o)\left(\frac{d\phi}{d\omega}\right)^2_{\omega_o} \end{aligned}$$

$$= \left(\frac{dA}{d\omega}\right)^2_{\omega_o} + \left(\frac{d\phi}{d\omega}\right)^2_{\omega_o}. \tag{2.35}$$

Note that we have utilized $H(j\omega_o) = -1$ in simplifying (2.35). We therefore have from (2.31)

$$|H_c(\omega_o + \Delta\omega)|^2 \approx \frac{1}{(\Delta\omega)^2 \left[\left(\frac{dA}{d\omega}\right)^2_{\omega_o} + \left(\frac{d\phi}{d\omega}\right)^2_{\omega_o}\right]}. \tag{2.36}$$

Because the higher the quality factor of the LC resonator of the oscillator, the better the frequency selectivity of the oscillator, as illustrated graphically in Fig. 2.6, the rate of the change of the magnitude and that of the phase of H_o quantified by $dA/d\omega$ and $d\phi/d\omega$, respectively jointly can therefore be used to portrait the degree of the frequency selectivity or equivalently the quality factor of the oscillator. Based on this observation, Herzel and Razavi introduced the following general definition of the quality factor of oscillators at the resonant frequency with an equal weight assigned to $dA/d\omega$ and $d\phi/d\omega$ [26]

$$Q(\omega_o) = \frac{\omega_o}{2}\sqrt{\left(\frac{dA}{d\omega}\right)^2_{\omega_o} + \left(\frac{d\phi}{d\omega}\right)^2_{\omega_o}}. \tag{2.37}$$

It is interesting to note that since at ω_o,

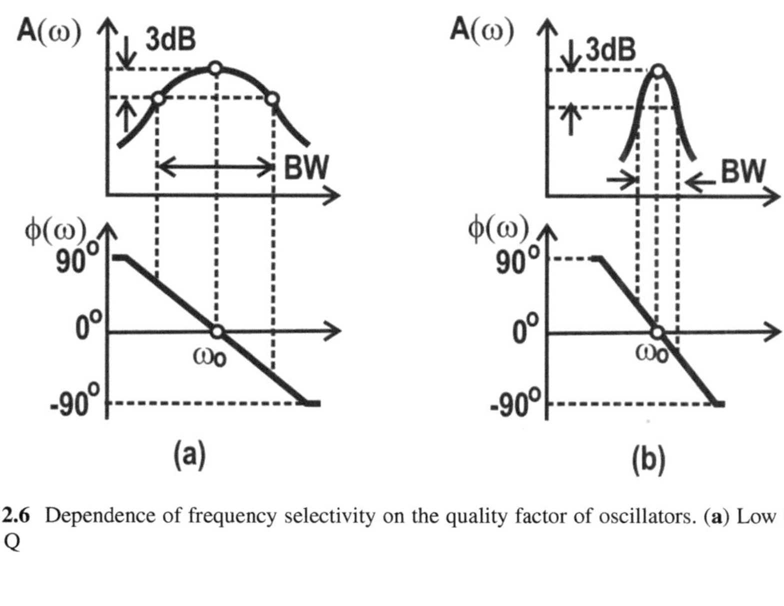

Fig. 2.6 Dependence of frequency selectivity on the quality factor of oscillators. (**a**) Low Q; (**b**) high Q

$$\left(\frac{dA}{d\omega}\right)_{\omega_o} \approx 0 \tag{2.38}$$

holds typically,[1] Eq. (2.37) can be simplified to the following widely used simplified expression of the quality factor of oscillators

$$Q(\omega_o) \approx \frac{\omega_o}{2}\left|\frac{d\phi}{d\omega}\right|_{\omega_o}. \tag{2.39}$$

Equation (2.39) agrees well with Fig. 2.6 as the higher $Q(\omega_o)$, the more larger the slope of $\phi(\omega)$, the better the frequency selectivity of the resonator.

The quality factor of the oscillator can also be measured using the width of the passband defined as the difference between the frequencies at which the drop of the magnitude of the open-loop transfer function from its value at ω_o is 3 dB. If the system is a second-order bandpass system whose transfer function is given by

$$H_o(s) = \frac{\frac{\omega_o}{Q}s}{s^2 + s\frac{\omega_o}{Q} + \omega_o^2}, \tag{2.40}$$

it can be shown that the two -3 dB frequencies of the system are given by [69]

$$\omega_1 = -\frac{\omega_o}{2Q} + \omega_o\sqrt{1 + \frac{1}{4Q^2}}, \tag{2.41}$$

and

$$\omega_2 = \frac{\omega_o}{2Q} + \omega_o\sqrt{1 + \frac{1}{4Q^2}}. \tag{2.42}$$

It follows from (2.41) and (2.42) that

$$\omega_1\omega_2 = \omega_o^2, \tag{2.43}$$

and the bandwidth of the system is given by

$$\Delta\omega_B = \omega_2 - \omega_1 = \frac{\omega_o}{Q}. \tag{2.44}$$

It is evident from (2.44) that the smaller the bandwidth, the higher the quality factor.

[1] Since $H_o = g_m Z_T = Ae^{j\phi}$, $|Z_T|$ needs to be symmetrical with respect to ω_o in order to have $\left(\frac{dA}{d\omega}\right)_{\omega_o} \approx 0$.

Let us now examine the power spectral density (PSD) of the output noise of harmonic oscillators. Rewrite (2.36) in terms of $Q(\omega_o)$

$$|H_c(\omega_o + \Delta\omega)|^2 \approx \frac{1}{4Q^2(\omega_o)} \left(\frac{\omega_o}{\Delta\omega}\right)^2. \quad (2.45)$$

For oscillators with a sufficiently large Q, the contribution of the sideband components of a broadband noise source to the output noise of the oscillator is negligibly small and can therefore be neglected safely. As a result, the output noise of the oscillator is mainly contributed by the noise of the broadband noise source that falls into the passband of the oscillator. The power spectral density of the output noise of the oscillator is given by

$$S_o(\omega_o) \approx \frac{1}{4Q^2(\omega_o)} \left(\frac{\omega_o}{\Delta\omega}\right)^2 S_{in}(\omega_o + \Delta\omega). \quad (2.46)$$

Equation (2.46) shows that the PSD of the output noise of the oscillator is inversely proportional to the square of the quality of the oscillator at ω_o. The higher the quality factor, the lower the PSD of the noise of the oscillator. It is also inversely proportional to the square of frequency displacement from ω_o. The larger the frequency displacement from ω_o, the lower the PSD. Since $S_o(\omega_o)$ is also proportional to $S_{in}(\omega_o)$, the PSD of harmonic oscillators due to thermal noise and that due to flicker noise source differ. If the input noise is the thermal noise of the channel current of a MOS transistor whose PSD is given by

$$S_{in}(\omega_o + \Delta\omega) = 4kT\gamma g_m, \quad (2.47)$$

where γ is a process-dependent constant, T is temperature in degrees Kelvin, and k is Boltzmann's constant or the input noise comes from the thermal noise (voltage) of a resistor whose PSD is given by

$$S_{in}(\omega_o + \Delta\omega) = 4kTR, \quad (2.48)$$

where R is the resistance of the resistor, Eq. (2.46) shows that the PSD of the output noise of the oscillator due to the thermal noise of the resistor or that of the channel current of the transistor is inversely proportional to the square of frequency displacement $(\Delta\omega)^2$.

In oscillators, the flicker noise of the tail current source transistor is up-converted to the vicinity of ω_o, as shown graphically in Fig. 2.7. We will detail the mechanism of the up-conversion of flicker noise in just a moment. Because flicker noise at frequency $\Delta\omega$ is up-converted to $\omega_o + \Delta\omega$, the power of the up-converted flicker noise at frequency $\omega_o + \Delta\omega$ is given by

$$S_{in}(\omega_o + \Delta\omega) = \frac{K_f I_{DS}}{\Delta\omega}, \quad (2.49)$$

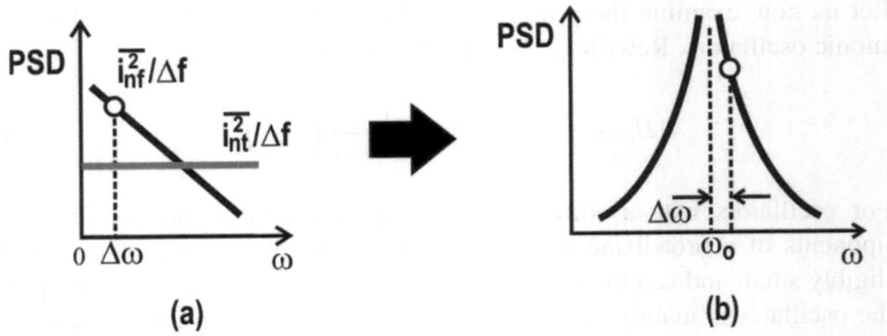

Fig. 2.7 Up-conversion of flicker noise in harmonic oscillators. i_{nt} and i_{nf} are the thermal and flicker noise currents of tail current source transistor, respectively. (**a**) Noise spectrum of transistor; (**b**) noise spectrum of oscillator

where K_f is a process-dependent constant and I_{DS} is the channel current of the transistor. It follows from (2.46) that the PSD of the output noise of the oscillator due to the flicker noise of the tail current source transistor is inversely proportional to $(\Delta\omega)^3$.

Figure 2.8d sketches the spectrum of the oscillator with the presence of both thermal and flicker noise. Since the power of flicker noise is concentrated in the neighborhood of ω_o and is significantly higher than that of thermal noise, the noise spectrum of a harmonic oscillator in the vicinity of its oscillation frequency is dominated by flicker noise. The intercept point at which the power of the flicker noise is the same as that of the thermal noise identifies the boundary beyond which thermal noise dominates. It is known as the corner frequency of flicker noise.

Since the power of flicker noise soars at low frequencies, the flicker noise of the tail current source transistor M3 is critical. One might argue that since M3 is biased in saturation and operated as a DC current source, the power of its flicker noise, which is mainly concentrated at low frequencies with corner frequency in a low MHz range [23], should not be of a concern as the frequency of oscillators used in RF applications is typically much higher. Due to the up-conversion of the flicker noise of the tail current source arising from the periodic sampling of the flicker noise by the oscillator, the flicker noise component of the tail current source at $\omega \ll \omega_o$ appears as a noise component at $\omega_o+\omega$ in the spectrum of the oscillator. To illustrate this, consider the LC oscillator shown in Fig. 2.8a. To simplify analysis, we neglect the noise generated by M1 and M2, as well as the thermal noise of M3, and only concentrate on the flicker noise of M3. We further assume that M1 and M2 operate in an ON–OFF mode with a 50% duty cycle. The operation of M1 and M2 can be depicted mathematically using switching function $s(t)$ and its complementary $\overline{s(t)}$ that are periodic with period T and a 50% duty cycle, as shown in Fig. 2.8c. Since $s(t)$ is periodic, it can be represented by Fourier series

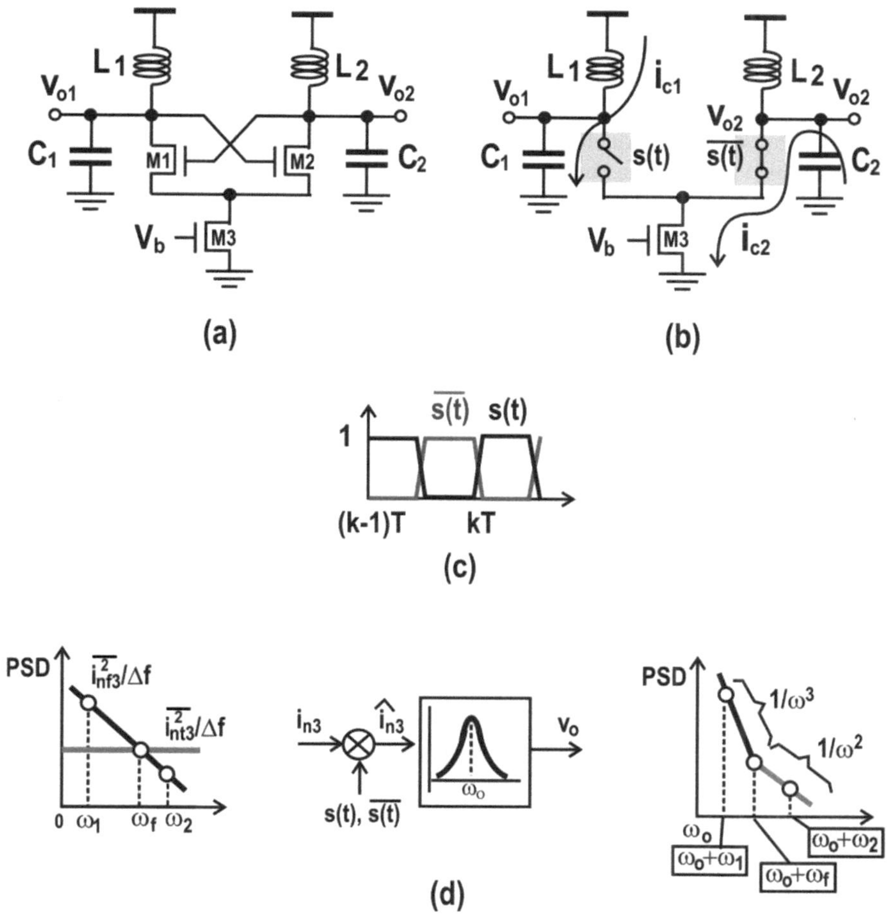

Fig. 2.8 Up-conversion of flicker noise in LC harmonic oscillators. ω_f is the corner frequency of flicker noise, and i_{nt3} and i_{nf3} are the thermal and flicker noise currents of M3, respectively

$$s(t) = \sum_{k=-\infty}^{\infty} S_k e^{jk\omega_o t}, \tag{2.50}$$

where

$$\begin{aligned} S_k &= \frac{1}{T_o}\int_0^T s(t)e^{-jk\omega_o t}dt \\ &= \frac{1}{2}e^{j\frac{k\pi}{2}}\,\text{sinc}\left(\frac{k\pi}{2}\right), \end{aligned} \tag{2.51}$$

and $\omega_o = 2\pi/T$. For the case shown in Fig. 2.8b where M1 is OFF and M2 is ON, C_1 is charged by V_{DD} via inductor L_1. C_2 is charged by V_{DD} via inductor L_2 and discharged to ground via M3 simultaneously. In order to allow C_2 to discharge, the discharging current provided by M3 must be larger than the charging current provided by V_{DD} via inductor L_2. For simplicity, we neglect the charging current provided by V_{DD}. Let the current of M3 be

$$i_{DS3} = I_{DS3} + i_{n3}, \tag{2.52}$$

where I_{DS3} is the DC channel current and i_{n3} is the noise of the channel current. The modulated discharge current of C_2 is given by

$$\begin{aligned} i_{c2} &= (I_{DS3} + i_{n3})\, s(t) \\ &= (I_{DS3} + i_{n3}) \sum_{k=-\infty}^{\infty} S_k e^{jk\omega_o t}. \end{aligned} \tag{2.53}$$

The voltage across C_2 at the end of the discharge interval is given by

$$v_{c2} = v_{c2}(0^-) - \frac{1}{C_2} \sum_{k=-\infty}^{\infty} \int_0^{T/2} (I_{DS3} + i_{n3})\, S_k e^{jk\omega_o t} dt. \tag{2.54}$$

Consider the noise of tail current source transistor M3. At ω_1 in Fig. 2.8d, since $\overline{i^2_{nf3}} \gg \overline{i^2_{nt3}}$ where $\overline{i^2_{nf3}}$ and $\overline{i^2_{nt3}}$ are the power of the flicker and thermal noise of M3, respectively, only the flicker noise is considered, i.e., $\overline{i^2_{n3}} \approx \overline{i^2_{nf3}}$. At ω_2, only the thermal noise needs to be considered, i.e., $\overline{i^2_{n3}} \approx \overline{i^2_{nt3}}$. The noise current of tail current source transistor M3 at ω_1 can be represented by

$$i_{n3} \approx i_{nf3} e^{j\omega_1 t}. \tag{2.55}$$

When modulated by $s(t)$, the modulated noise current will be given by

$$\hat{i}_{n3} \approx i_{nf3} \sum_{k=-\infty}^{\infty} S_k e^{j(\omega_1 + k\omega_o)t}. \tag{2.56}$$

The output of the oscillator is the voltage drop across the LC resonator. Although the periodic modulation of the noise current of the tail current source by the cross-coupled transistors gives rise to an infinite number of frequency components, only those that fall into the passband of the LC resonator will appear at the output of the oscillator whole all others are suppressed by the resonator. To demonstrate this further, let us examine both up-aliasing and down-aliasing cases:

1. *Up-aliasing*: Consider flicker noise component at $\omega_1 \ll \omega_f$. The modulated noise current is given by (2.56). Since $\omega_1 \ll \omega_f$, $\omega_1 + \omega_o \approx \omega_o$. Only the frequency component of i_{nf3} at $\omega_o + \omega_1$ will make its way to the output of the oscillator while those frequency components of i_{nf3} at $\omega_1 + k\omega_o$ with $k \neq 1$ will be attenuated by the band-pass characteristics of the LC resonator. The appearance of the flicker noise of the tail current source in the vicinity of ω_o at the output of the oscillator is known as the up-conversion of the flicker noise of the tail current source transistor. Since the power of the flicker noise of M3 at low frequencies is much larger as compared with that of its thermal noise at the same frequencies, the flicker noise portion of the spectrum of the oscillator dominates the phase noise of the oscillator in the vicinity of ω_o, as shown in Fig. 2.8d. Note that although the thermal noise current of M3 at ω_1 is also up-converted, since it is much smaller as compared with the flicker noise at ω_1, its effect is not of a concern.

   Let us now consider the thermal noise at $\omega_f \ll \omega_2 \ll \omega_o$. The modulated noise current is given by

$$\hat{i}_{n3} \approx i_{nt3} \sum_{k=-\infty}^{\infty} S_k e^{j(\omega_2 + k\omega_o)t}. \tag{2.57}$$

   Since $\omega_2 \gg \omega_f$, ω_2 well exceeds the bandwidth of the LC resonator, $\hat{i}_{n3}$ will be largely attenuated by the band-pass characteristics of the resonator.

2. *Down-aliasing*: Let us consider the thermal noise of the tail current source at frequencies $\omega_1 + m\omega_o$ with $m = 1, 2, 3, \ldots$. It can be depicted by

$$i_{n3} = I_{nt3} e^{j(\omega_1 + m\omega_o)t}. \tag{2.58}$$

   The modulated current is given by

$$\hat{i}_{n3} = I_{nt3} \sum_{m,k=-\infty}^{\infty} S_k e^{j[\omega_1 + (m+k)\omega_o]t}. \tag{2.59}$$

   The frequency component of $\hat{i}_{nt3}$ at $\omega_1 + \omega_o$ where $m + k = 1$ falls into the pass-band of the LC resonator and will make its way to the output of the oscillator without attenuation. In other words, thermal noise components at high frequencies will contribute to the output noise of the oscillator. This is known as aliasing or fold-over effect and is a unique characteristic of periodically time-varying circuits. Since the high-order components of $s(t)$ is much smaller as compared with its fundamental component, the contribution of the thermal noise at high frequencies to the phase noise of the oscillator is rather marginal.

The preceding analysis demonstrates that as far as the noise of the oscillator is concerned, the up-conversion of the flicker noise of the tail current source transistor of the oscillator is of the most critical concern.

2.4 Modeling of Harmonic Oscillators

A convenient and effective way to study the injection-locking process of a harmonic oscillator is to model the oscillator using a nonlinear system with a nonlinear block $y = f(x)$ characterizing the nonlinearities of the oscillator such as the $i \sim v$ characteristics of MOSFETs and a band-pass filter with transfer function $H(j\omega)$ that depicts the band-pass characteristics of LC resonator, as shown in Fig. 2.9 [63, 83]. The nonlinear block of the oscillator is memoryless such that its behavior can be characterized using power series

$$y = \sum_{k=1}^{\infty} a_k x^k, \tag{2.60}$$

where a_k is constant. Since the frequency selection of harmonic oscillators is achieved using a LC resonator functioning as a high-Q band-pass filter, the band-pass filter of the oscillators has its transfer function given by [68, 110]

$$H(j\omega) = \frac{H_o}{1 + j2Q\left(\frac{\omega-\omega_o}{\omega_o}\right)}, \tag{2.61}$$

where ω_o is the center frequency of the passband, e.g., the resonant frequency of the LC resonator, H_o is the gain of the filter at ω_o, and Q is the quality factor of the filter. It is straightforward to show that at

$$\omega_{-3dB_{1,2}} = \left(1 \pm \frac{1}{2Q}\right)\omega_o. \tag{2.62}$$

We therefore have

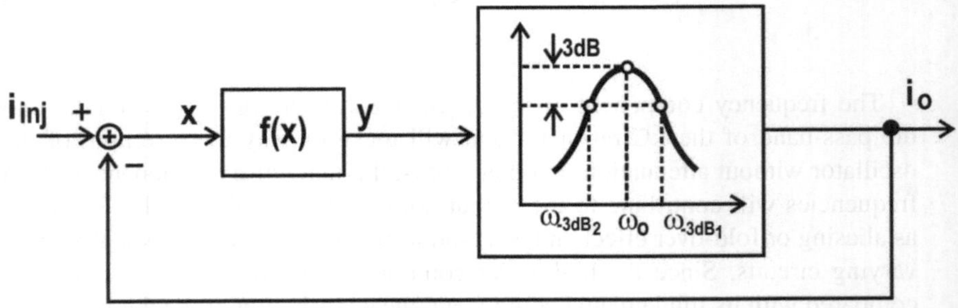

Fig. 2.9 Harmonic oscillators are represented by a negative feedback system with a nonlinear memoryless block $y = f(x)$ and a band-pass filter $H(s)$ in the forward path

$$|H(j\omega_{-3dB_{1,2}})| = \frac{H_o}{\sqrt{2}}. \tag{2.63}$$

The bandwidth of the band-pass filter, denoted by $\Delta\omega_B$, is obtained from

$$\Delta\omega_B = \omega_{-3dB_1} - \omega_{-3dB_2} = \frac{\omega_o}{Q}. \tag{2.64}$$

Let us first examine a harmonic oscillator without injection. To facilitate analysis, we assume that the quality factor of the band-pass filter of the oscillator is sufficiently high such that only those frequency components that fall into the passband of the filter pass through the filter. Those that are outside the passband of the filter are completely suppressed by the filter. Let the output of the filter be a sinusoid at ω

$$i_o = I_o \cos(\omega t) = \frac{I_o}{2}\left(e^{j\omega t} + e^{-j\omega t}\right). \tag{2.65}$$

Further, let the transfer characteristics of the nonlinear block of the oscillator be depicted by the following third-order Taylor series

$$y = a_1 x + a_2 x^2 + a_3 x^3, \tag{2.66}$$

where a_1, a_2, and a_3 are constants. Since $i_{inj} = 0$, i.e., no injection, the input of the nonlinear block is given by

$$x = -i_o. \tag{2.67}$$

We want to find out whether the nonlinear block of the oscillator generates any frequency component that is different from the passband center frequency ω_o of the band-pass filter at the output of the oscillator or not. The output of the nonlinear block is given by

$$\begin{aligned} y = &-\frac{a_1 I_o}{2}\left(e^{j\omega t} + e^{-j\omega t}\right) + \frac{a_2 I_o^2}{4}\left(e^{j\omega t} + e^{-j\omega t}\right)^2 \\ &-\frac{a_3 I_o^3}{8}\left(e^{j\omega t} + e^{-j\omega t}\right)^3. \end{aligned} \tag{2.68}$$

Among the frequency components of the output of the nonlinear block, only those whose frequency is located in the passband of the filter pass through the filter and reach the output of the oscillator. These frequency components are:

(a) $-\left(\frac{a_1 I_o}{2}\right) e^{j\omega t}$,

(b) $-\left(\frac{3a_3 I_o^3}{8}\right) e^{j\omega t}$.

All other frequency components are suppressed by the filter. We therefore have the following phasor relation at frequency ω

$$\frac{I_o}{2} = -\frac{H_o}{1 + j2Q\left(\frac{\omega-\omega_o}{\omega_o}\right)}\left(\frac{a_1 I_o}{2} + \frac{3a_3 I_o^3}{8}\right). \tag{2.69}$$

Let us first consider a special case where $a_3 \ll a_1$, i.e., the effect of the nonlinear characteristics of the nonlinear block is negligible, Eq. (2.69) in this case is simplified to

$$\frac{I_o}{2} \approx -\frac{H_o}{1 + j2Q\left(\frac{\omega-\omega_o}{\omega_o}\right)}\frac{a_1 I_o}{2}. \tag{2.70}$$

Separating the real and imaginary parts of (2.70) yields

$$1 + a_1 H_o = 0 \tag{2.71}$$

and

$$2Q\left(\frac{\omega - \omega_o}{\omega_o}\right) = 0 \tag{2.72}$$

It is seen from (2.72) that the frequency of the oscillator is the passband center frequency of the band-pass filter. Equation (2.71) sets the constraint imposed on the gain of the filter at its passband center frequency. This constraint is the result of Barkhausen magnitude criterion as (2.71) is the denominator of the closed-loop transfer function of the oscillator when the effect of the nonlinearity is neglected.

Let us now consider a more general case where a_3 cannot be neglected. We obtain the frequency of the output of the oscillator by separating the real and imaginary parts of (2.69)

$$1 + \left(a_1 + \frac{3a_3}{4} I_o^2\right) H_o = 0 \tag{2.73}$$

and

$$2Q\left(\frac{\omega - \omega_o}{\omega_o}\right) = 0. \tag{2.74}$$

It is evident from (2.74) that the frequency of the output of the oscillator is still the center frequency of the band-pass filter regardless of the degree of the nonlinearity of the oscillator. Equation (2.73) sets the constraint on the gain of the filter at its passband center frequency in order to satisfy Barkhausen magnitude criterion. A comparison of (2.71) and (2.73) reveals when a_3 cannot be neglected, since $a_3 < 0$

typically due to the gain compression nature of the nonlinearities of MOSFETs, the gain of the filter at ω_o will need to be larger in order to satisfy Barkhausen magnitude criterion.

The preceding analysis demonstrates that the oscillation frequency of harmonic oscillators is the center frequency of the band-pass filter. It is independent of the degree of the nonlinearity of the nonlinear block of the oscillators. The degree of the nonlinearity of the nonlinear block of the oscillator only affects the gain of the band-pass filter at the passband center frequency required to start and sustain oscillation.

2.5 First-Harmonic Injection

In this section, we investigate the lock range of injection-locked harmonic oscillators with the frequency of injection signals located in the vicinity of the free-running frequency of harmonic oscillators. We term it first-harmonic injection. The frequency of the injection signal can also be either a superharmonic or a subharmonic of the free-running frequency of the oscillator. They are termed superharmonic injection and subharmonic injection, respectively, and will be investigated in the next two sections. The lock range of harmonic oscillators in first-harmonic injection will be investigated from both a nonlinear or linear perspectives. The former takes into account the nonlinear characteristics of harmonic oscillators while the latter treats injection-locked oscillators as linear systems. The nonlinear approach provides the critical information of the locking process of harmonic oscillators while the linear approach yields the lock range of the oscillators without the need for the prior knowledge of the locking process of the oscillators.

2.5.1 *Nonlinear System Approach*

Consider a harmonic oscillator with the presence of a sinusoidal injection signal i_{inj}

$$\begin{aligned} i_{inj} &= I_{inj}\cos(\omega_{inj}t+\phi_{inj}) \\ &= \frac{I_{inj}}{2}\left(e^{j\omega_{inj}t}e^{j\phi_{inj}}+e^{-j\omega_{inj}t}e^{-j\phi_{inj}}\right). \end{aligned} \tag{2.75}$$

The block diagram of the oscillator is shown in Fig. 2.9 with the characteristics of the nonlinear block depicted by (2.66). Let the output of the oscillator in the lock state be a sinusoid of frequency ω

$$\begin{aligned} i_o &= I_o\cos(\omega t) \\ &= \frac{I_o}{2}\left(e^{j\omega t}+e^{-j\omega t}\right). \end{aligned} \tag{2.76}$$

Note that the phase of the injection signal ϕ_{inj} is measured with reference to the phase of i_o, which is chosen to be the reference. Since the output of the oscillator is the output of the band-pass filter, ω is in the passband of the band-pass filter. In order to achieve lock, the frequency of the injection signal should not be too far away from the free-running frequency of the oscillator, i.e., the passband center frequency of the band-pass filter. It is reasonable to assume that ω_{inj} fall into the passband of the band-pass filter. The input of the nonlinear block of the oscillator with the presence of the injection signal is given by

$$\begin{aligned} x &= i_{inj} - i_o \\ &= \frac{I_{inj}}{2} e^{j\omega_{inj}t} e^{j\phi_{inj}} + \frac{I_{inj}}{2} e^{-j\omega_{inj}t} e^{-j\phi_{inj}} - \frac{I_o}{2} e^{j\omega t} - \frac{I_o}{2} e^{-j\omega t}. \end{aligned} \tag{2.77}$$

The output of the nonlinear block is obtained from

$$\begin{aligned} y = a_1 &\left(\frac{I_{inj}}{2} e^{j\omega_{inj}t} e^{j\phi_{inj}} + \frac{I_{inj}}{2} e^{-j\omega_{inj}t} e^{-j\phi_{inj}} - \frac{I_o}{2} e^{j\omega t} - \frac{I_o}{2} e^{-j\omega t} \right) \\ +a_2 &\left(\frac{I_{inj}}{2} e^{j\omega_{inj}t} e^{j\phi_{inj}} + \frac{I_{inj}}{2} e^{-j\omega_{inj}t} e^{-j\phi_{inj}} - \frac{I_o}{2} e^{j\omega t} - \frac{I_o}{2} e^{-j\omega t} \right)^2 \\ +a_3 &\left(\frac{I_{inj}}{2} e^{j\omega_{inj}t} e^{j\phi_{inj}} + \frac{I_{inj}}{2} e^{-j\omega_{inj}t} e^{-j\phi_{inj}} - \frac{I_o}{2} e^{j\omega t} - \frac{I_o}{2} e^{-j\omega t} \right)^3. \end{aligned} \tag{2.78}$$

The output of the nonlinear block contains various frequency components. Only those frequency components that fall into the passband of the band-pass filter pass through the filter and reach the output of the oscillator. All other frequency components are suppressed by the band-pass filter. The frequency components of the output of the nonlinear block that reside in the passband of the bandpass filter are:

(a) $\frac{a_1}{2} I_{inj} e^{j\phi_{inj}} e^{j\omega_{inj}t}$,
(b) $-\frac{a_1}{2} I_o e^{j\omega t}$,
(c) $\frac{3a_3}{8} I_{inj}^3 e^{j\phi_{inj}} e^{j\omega_{inj}t}$,
(d) $-\frac{3a_3}{8} I_{inj}^2 I_o e^{j2\phi_{inj}} e^{j(2\omega_{inj}-\omega)t}$,
(e) $\frac{3a_3}{8} I_{inj} I_o^2 e^{-j\phi_{inj}} e^{j(2\omega-\omega_{inj})t}$,
(f) $-\frac{3a_3}{8} I_o^3 e^{j\omega t}$.

Because when the lock state is established, the oscillator will oscillate at the frequency of the injection signal, e.g., $\omega = \omega_{inj}$. The exponential terms in the above expressions will become identical. As a result, the following phasor relation at ω_{inj} exists

$$\frac{I_o}{2} = \frac{H_o}{1 + j2Q\left(\frac{\omega_{inj}-\omega_o}{\omega_o}\right)} \tag{2.79}$$
$$\times \left[\frac{a_1}{2}\left(I_{inj}e^{j\phi_{inj}} - I_o\right) + \frac{3a_3}{8}(I_{inj}^3 e^{j\phi_{inj}} - I_{inj}^2 I_o e^{j2\phi_{inj}} + I_{inj} I_o^2 e^{-j\phi_{inj}} - I_o^3)\right].$$

A comparison of (2.69) for harmonic oscillators without injection and (2.79) for harmonic oscillators in the lock state reveals both similarities and differences between them. It is evident that (2.79) will be simplified to (2.69) when $I_{inj} = 0$. Since (2.79) is valid only when the oscillator is in the lock state, $\omega_{inj} - \omega_o$ obtained from (2.79) gives the maximum difference between the frequency of the injection signal and the natural frequency of the oscillator and is therefore the lock range of the oscillator

$$\Delta\omega_L = \omega_{inj} - \omega_o. \tag{2.80}$$

Clearly lock range $\Delta\omega_L$ is a function of injection strength I_{inj}, injection phase ϕ_{inj}, and the degree of the nonlinearity of the nonlinear block of the oscillator under injection.

Let us first consider the special case where the intermodulation terms in (2.79) can be neglected. This is the case when $a_3 \ll a_1$. Equation (2.79) in this case is simplified to

$$\frac{I_o}{2} \approx \frac{H_o}{1 + j2Q\left(\frac{\omega_{inj}-\omega_o}{\omega_o}\right)} \frac{a_1}{2}\left(I_{inj}e^{j\phi_{inj}} - I_o\right). \tag{2.81}$$

Separating the real and imaginary parts of (2.81) yields

$$1 + a_1 H_o - a_1 H_o \frac{I_{inj}}{I_o} \cos\phi_{inj} = 0, \tag{2.82}$$

and

$$\Delta\omega_L = \frac{\omega_o}{2Q}\frac{I_{inj}}{I_o}(a_1 H_o)\sin\phi_{inj}. \tag{2.83}$$

Equation (2.83) shows that $\Delta\omega_L$ will be maximized when

$$\phi_{inj}^* = \frac{\pi}{2}. \tag{2.84}$$

The maximum lock range is given by

$$\Delta\omega_{L,max} = \frac{\omega_o}{2Q}\frac{I_{inj}}{I_o}(a_1 H_o). \tag{2.85}$$

At the maximum lock range where $\phi^*_{inj} = \frac{\pi}{2}$, Eq. (2.82) is simplified to (2.71), which is the constraint imposed on the gain of the band-pass filter at its passband center frequency in order to satisfy Barkhausen magnitude criterion. With

$$1 + a_1 H_o = 0 \tag{2.86}$$

at the maximum lock range, Eq. (2.85) becomes

$$\Delta\omega_{L,max} = -\frac{\omega_o}{2Q}\frac{I_{inj}}{I_o}. \tag{2.87}$$

Equation (2.87) reveals that the maximum lock range of harmonic oscillators is directly proportional to injection strength $\frac{I_{inj}}{I_o}$ and inversely proportional to the quality factor of Q of the oscillator.

The fact that the lock range of harmonic oscillators will be maximized when $\phi_{inj} = \frac{\pi}{2}$ bears a strong resemblance to the maximum phase noise generation of oscillators. To demonstrate this, since

$$i_o = I_o \cos(\omega t), \tag{2.88}$$

$\frac{di_o}{dt}$ will peak when

$$\omega t = (2n+1)\frac{\pi}{2} \tag{2.89}$$

and vanish when

$$\omega t = n\pi, \tag{2.90}$$

where $n = 0, \pm 1, \pm 2, \ldots$. The injection of i_{inj} with a phase $\phi_{inj} = \frac{\pi}{2}$ is equivalent to injecting i_{inj} at the threshold-crossing of i_o where $\frac{di_o}{dt}$ is maximized. On the other hand, the injection of i_{inj} with a phase $\phi_{inj} = 0$ is equivalent to injecting i_{inj} at the peak of i_o where $\frac{di_o}{dt}$ vanishes, as shown in Fig. 2.10. It is well understood that the effect of the perturbation of a noise signal on the phase noise of an oscillator is maximized if the noise signal is injected at the threshold-crossing of the output

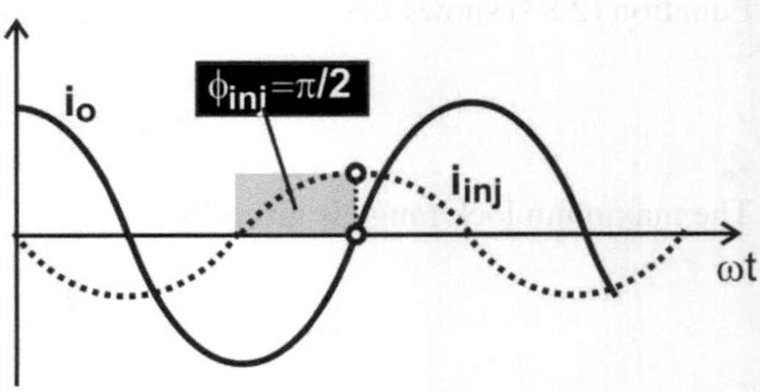

Fig. 2.10 First-harmonic injection of harmonic oscillators with i_{inj} leading i_o by $\pi/2$

of the oscillator and is minimized if it is injected at the peak of the output of the oscillator [25]. The preceding results echo well with the phase noise behavior of the oscillator.

Let us now consider a more general case where the intermodulation terms in (2.79) cannot be neglected. This is the case when the effect of the nonlinearity is accounted for. We follow the same approach as that for injection-locked harmonic oscillators with a_3 neglected. Specifically, we separate the real and imaginary parts of (2.79), which yield

$$I_o = a_1 H_o I_{inj} \cos\phi_{inj} - a_1 H_o I_o + \frac{3a_3}{4}\left(I_{inj}^3 \cos\phi_{inj} - I_{inj}^2 I_o \cos 2\phi_{inj} + I_{inj} I_o^2 \cos\phi_{inj} - I_o^3\right), \tag{2.91}$$

and

$$I_o\left[2Q\left(\frac{\omega_{inj}-\omega}{\omega_o}\right)\right] = a_1 H_o I_{inj} \sin\phi_{inj} + \frac{3a_3}{4}\left(I_{inj}^3 \sin\phi_{inj} - I_{inj}^2 I_o \sin 2\phi_{inj} - I_{inj} I_o^2 \sin\phi_{inj}\right). \tag{2.92}$$

Since $I_{inj} \ll I_o$ typically,

$$I_{inj}^3,\ I_{inj} I_o,\ I_{inj} I_o^2 \ll I_o^3 \tag{2.93}$$

holds. As a result, Eq. (2.91) can be simplified by dropping the terms associated with $I_{inj}^3 \cos\phi_{inj}$, $I_{inj}^2 I_o \cos 2\phi_{inj}$, and $I_{inj} I_o^2 \cos\phi_{inj}$. The result is given by

$$1 + a_1 H_o - a_1 H_o \frac{I_{inj}}{I_o} \cos\phi_{inj} + \frac{3a_3}{4} H_o I_o^2 = 0. \tag{2.94}$$

Similarly, Eq. (2.92) can be simplified by dropping the terms associated with $I_{inj}^3 \sin\phi_{inj}$ and $I_{inj}^2 I_o \sin 2\phi_{inj}$. The result is given by

$$\Delta\omega_L = \frac{\omega_o}{2Q}\frac{I_{inj}}{I_o}(a_1 H_o)\left(1 - \frac{3a_3}{4a_1} I_o^2\right) \sin\phi_{inj}. \tag{2.95}$$

We comment on the preceding results:

- A comparison of (2.82) with (2.94) reveals that the additional term in (2.94) is caused by the nonlinearity of the oscillator and is a function of both the degree of the third-order nonlinearity signified by a_3 and the output of the oscillator. The larger a_3 or I_o, the more severe the effect of the nonlinearity.

- A similar observation is made in comparison of (2.95) with (2.83). Since the nonlinear characteristics of MOSFETs is generally characterized with gain compression, $a_3 < 0$ holds typically. As a result, the lock range of harmonic oscillators with the impact of nonlinearity considered is larger as compared with that without. In other words, increasing the degree of the nonlinearity of harmonic oscillators will increase lock range. This characteristic is important as it provides designers with a much needed theoretical basis upon which new design techniques that increase the lock range of harmonic oscillators can be developed. We will come back to this in a great detail in the next chapter where design techniques to increase the lock range of harmonic oscillators are studied.

- The maximum lock range will occur when

$$\phi^* = \frac{\pi}{2}. \tag{2.96}$$

Equation (2.97) is the same as that without considering the effect of the nonlinearity revealing that the degree of the nonlinear characteristics of harmonic oscillators does not affect the optimal injection phase at which the maximum lock range is obtained. The maximum lock range is given by

$$\Delta\omega_L = \frac{\omega_o}{2Q}\frac{I_{inj}}{I_o}(a_1 H_o)\left(1 - \frac{3a_3}{4a_1}I_o^2\right). \tag{2.97}$$

- At the optimal injection phase, Eq. (2.94) becomes

$$1 + a_1 H_o + \frac{3a_3}{4}I_o^2 = 0. \tag{2.98}$$

Since $a_3 < 0$, H_o needs to be larger as compared with that without considering the impact of the nonlinearity of the oscillator in order to satisfy Barkhausen magnitude criterion. This is because the gain-compression nature of the nonlinear characteristics of MOSFETs will give rise to a smaller loop gain when the degree of the nonlinearity or the output of the oscillator increases. In this case, the band-pass filter must provide an additional gain in order to satisfy Barkhausen magnitude criterion.

2.5.2 *Linear System Approach*

The preceding nonlinear approach shows that although in the locking process of a harmonic oscillator multiple frequency components coexist, only those that fall into the passband of the resonator of the oscillator will appear at the output of the oscillator when the oscillator is locked to the injection signal. Since in the lock state, the frequency of the oscillator is the frequency of the input, both the input and output of an injection-locked harmonic oscillator have the same frequency, which is the frequency of the injection signal, regardless of the existence of other frequency components inside the oscillator. This observation suggests that an injection-locked harmonic oscillator can be treated as a linear system from a terminal behavior point of view. As a result, linear system analysis approaches can be employed to analyze the oscillator in the lock state conveniently. It should be emphasized that linear system approaches can only yield some of the wanted information of the oscillator, in particular, the lock range of the oscillator by analyzing the terminal behavior of the oscillator. It cannot gain the critical information of the locking process of the oscillator such as the generation of frequency components other than that of the injection signal, the mixing of frequency components inside the oscillator, and the manipulation, propagation, and diminishing process of frequency components at frequencies other than that of the injection signal before reaching the output nodes of the oscillator. Such information, however, is often pivotal to our understanding of the locking process of harmonic oscillators, in particular, the mechanism upon which the lock range is determined. We will come back this point in a great detail in the next chapter when techniques to boost the lock range of harmonic oscillators are studied.

Consider the harmonic oscillator shown in Fig. 2.11. At the resonant frequency of the LC resonator, the tuned amplifier provides a $-180°$ phase shift. In order to satisfy Barkhausen phase criterion, an additional $-180°$ phase shift is needed. This additional phase shift is provided by the unity-gain inverting amplifier inserted between the drain and gate of the MOS transistor. Let i_{inj} be the current injected into the drain of the transistor. When the oscillator is locked to the frequency of the injection signal, its oscillation frequency will be the same as the frequency of the injection signal ω_{inj}. In other words, only one frequency, which is the frequency of the injection signal, will exist in the system when the oscillator is locked to the injection signal.

Let Z_T be the impedance looking into the resonator. Write Kirchhoff's Current Law at the drain of the transistor at ω_{inj}

$$I_{osc} + I_T = I_{inj}. \tag{2.99}$$

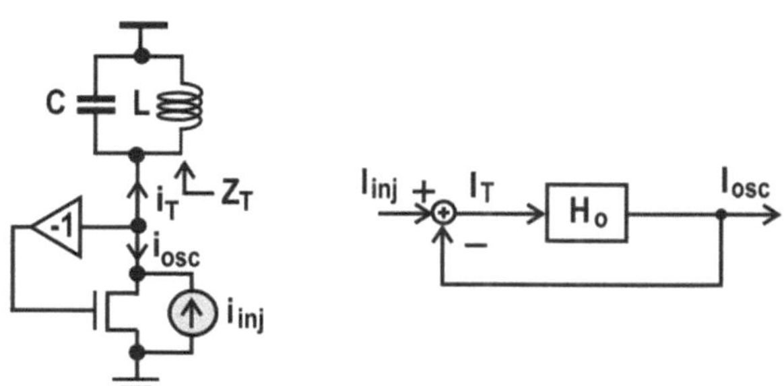

Fig. 2.11 Left: Injection-locked harmonic oscillator. Right: Negative feedback system depicting the injection-locked harmonic oscillator

Note that all variables here are their AC components. The gate voltage of the transistor is given by

$$V_g = -Z_T I_T. \tag{2.100}$$

We have

$$I_{osc} = -g_m Z_T I_T, \tag{2.101}$$

where g_m is the transconductance of the transistor. Note that the preceding equations are valid at ω_{inj}. Equation (2.99) can be represented by the linear negative feedback system shown in Fig. 2.11 with

$$H_o = -g_m Z_T. \tag{2.102}$$

Note that the linear negative feedback system is valid at ω_{inj} only. This is because we have assumed that the transistor behaves as a linear transconductance rather than a nonlinear transconductor. As a result, the only frequency that exists in the injection-locked oscillator is ω_{inj}. When the injection signal is absent, the oscillator will be an autonomous system with its oscillation frequency given by $\omega_o = 1/\sqrt{LC}$. Consider that a sinusoidal current I_{inj} of frequency $\omega_{inj} = \omega_o + \Delta\omega$ with $\Delta\omega \ll \omega$ is injected into the drain of the transistor, as shown in Fig. 2.12. If $\Delta\omega$ is within the lock range of the oscillator, the oscillator will lock to the injection signal and oscillate at $\omega_{inj} = \omega_o + \Delta\omega$. Since $\omega_{inj} \neq \omega_o$, the impedance of the resonator at $\omega_o + \Delta\omega$, denoted by $Z_T(\omega_o + \Delta\omega)$, will deviate from its value at the natural oscillation frequency ω_o, denoted by $Z_T(\omega_o)$, of the oscillator, as illustrated graphically in Fig. 2.12. We can write

$$Z_T(\omega_o + \Delta\omega) = Z_T(\omega_o) - \Delta Z_T, \tag{2.103}$$

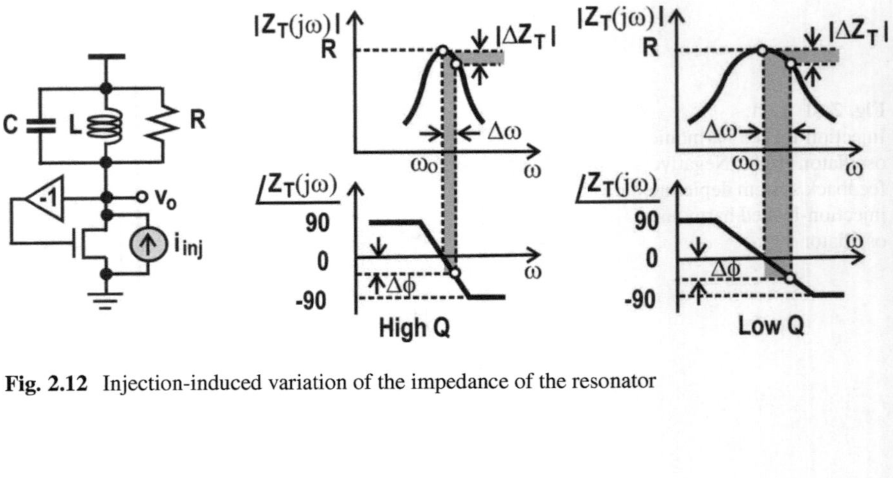

Fig. 2.12 Injection-induced variation of the impedance of the resonator

where ΔZ_T denotes the change of the impedance of the resonator induced by the injection signal. Since the injection-locked oscillator oscillates at $\omega_{inj} = \omega_o + \Delta\omega$, Barkhausen criteria are satisfied at ω_{inj}. This observation suggests that the injection-locked oscillator can be treated as an autonomous system, i.e., without injection, with its "natural" oscillation frequency $\omega_o + \Delta\omega$ or equivalently the impedance of the resonator is changed from $Z_T(\omega_o)$ to $Z_T(\omega_o + \Delta\omega)$. For the same injection signal, since the injection-induced impedance variation

$$\Delta Z_T = |\Delta Z_T| e^{j\Delta\phi} \tag{2.104}$$

is the same, the larger the quality factor of the resonator, the smaller the frequency displacement $\Delta\omega$, as illustrated graphically in Fig. 2.12. As lock range $\Delta\omega$ is the maximum frequency displacement induced by the injection signal, it is thus inversely proportional to the quality factor of the LC resonator.

Now let us consider the negative feedback system in Fig. 2.11. When the oscillator is locked to the injection signal, it will oscillate at $\omega_{inj} = \omega_o + \Delta\omega$ with its output current I_{osc} obtained from

$$I_{osc}(\omega_o + \Delta\omega) = \frac{H_o(\omega_o + \Delta\omega)}{1 + H_o(\omega_o + \Delta\omega)} I_{inj}(\omega_o + \Delta\omega). \tag{2.105}$$

We are only interested in the behavior of the oscillator in the lock state rather than the locking process. This is because although the locking process is a nonlinear process and multiple frequency components coexist, only a single frequency, i.e., the frequency of the injection signal, will exist at the output of the oscillator once the oscillator is locked to the injection signal [87].

If the strength of the injection signal is weak, $\Delta\omega \ll \omega_o$ will hold. Since

$$H_o(\omega_o + \Delta\omega) = -g_m Z_T(\omega_o + \Delta\omega), \tag{2.106}$$

the first-order approximation can be employed in this case to estimate $H_o(\omega_o + \Delta\omega)$

$$H_o(\omega_o + \Delta\omega) \approx H_o(\omega_o) + \left[\frac{dH_o}{d\omega}\right]_{\omega_o} \Delta\omega. \tag{2.107}$$

Equation (2.105) becomes

$$I_{osc}(\omega_{inj}) \approx \frac{H_o(\omega_o + \Delta\omega)}{\left(\frac{dH_o}{d\omega}\right)_{\omega_o} \Delta\omega} I_{inj}(\omega_{inj}). \tag{2.108}$$

Focusing on the amplitude only, we have

$$\left|I_{osc}(\omega_{inj})\right| \approx \frac{|H_o(\omega_o + \Delta\omega)|}{\left|\frac{dH_o}{d\omega}\right|_{\omega_o}|\Delta\omega|}|I_{inj}\omega_{inj})|. \tag{2.109}$$

Making use of the definition of the quality factor defined earlier in (2.37), we arrive at

$$\left|\frac{dH_o}{d\omega}\right|_{\omega_o} = \frac{2Q}{\omega_o}. \tag{2.110}$$

Substituting (2.110) into (2.109) yields

$$|I_{osc}(\omega_{inj})| \approx \frac{|H_o(\omega_o + \Delta\omega)|}{\frac{2Q}{\omega_o}|\Delta\omega|}|I_{inj}(\omega_{inj})|. \tag{2.111}$$

Since

$$I_{osc}(\omega_{inj}) = H_o(\omega_o + \Delta\omega)I_T(\omega_{inj}), \tag{2.112}$$

we have

$$|\Delta\omega| \approx \frac{\omega_o}{2Q}\left|\frac{I_{inj}(\omega_{inj})}{I_T(\omega_{inj})}\right|. \tag{2.113}$$

Because $|\Delta\omega|$ quantifies the maximum deviation of the frequency of the oscillator from its natural frequency ω_o under the condition that the oscillator is locked to the injection signal, $|\Delta\omega|$ is the lock range of the oscillator. Since I_{inj} is small, $I_T \approx I_{osc}$ holds, Eq. (2.113) can therefore be written as

$$|\Delta\omega| \approx \frac{\omega_o}{2Q}\left|\frac{I_{inj}(\omega_{inj})}{I_{osc}(\omega_{inj})}\right|. \tag{2.114}$$

Equation (2.114) reveals that the lock range of the injection-locked harmonic oscillator is directly proportional to the relative strength of the injection signal and inversely proportional to the quality factor of the oscillator.

For harmonic oscillators with an active inductor resonator, since the inductance of active inductors varies with the transconductance of the transconductors constituting the active inductors, injection signals in this case assert an additional impact on Z_T subsequently the lock range. Let us investigate this in detail. It was pointed out earlier that an injection-locked harmonic oscillator can be treated as an autonomous system with the impact of the injection signal represented by the variation of the impedance of the resonator. To quantify the relation between the injection signal and the variation of the impedance of the resonator, we consider Fig. 2.13a where injection signal I_{inj} at frequency $\omega_o + \Delta\omega$ is injected into node 2 of the harmonic oscillator. The frequency and strength of the injection signal are properly chosen

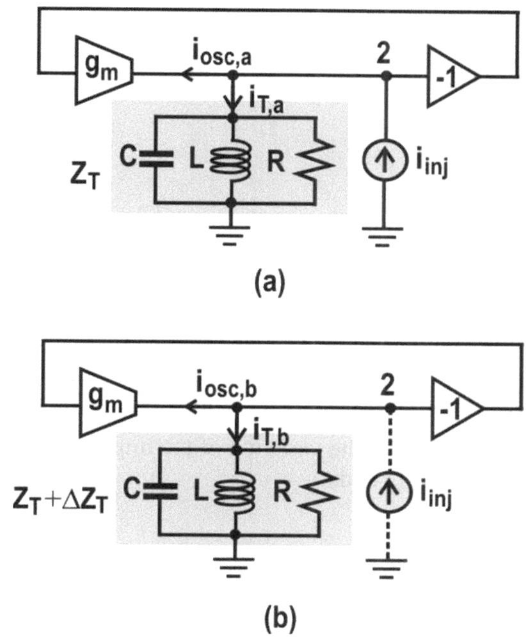

Fig. 2.13 (**a**) Injection-locked harmonic oscillators. (**b**) The equivalent circuit of injection-locked harmonic oscillators with injection signal removed and the effect of the injection signal represented by the variation of the impedance of the resonator

such that the oscillator is locked to the injection signal and oscillates at the frequency of the injection signal $\omega_o + \Delta\omega$. In Fig. 2.13b, the impact of the injection signal I_{inj} on the oscillator is represented by the variation of the impedance of the resonator and the injection-locked oscillator is treated as an autonomous system.

The two systems in Fig. 2.13 are said to be equivalent if both the voltage and current at node 2 of the two networks are identical, e.g., $V_{2,a} = V_{2,b}$ and $I_{osc,a} = I_{osc,b}$ are satisfied simultaneously where $V_{2,a}$ and $V_{2,b}$ are the voltage of node 2 in (a) and (b), respectively, and $I_{osc,a} = I_{osc,b}$ are defined explicitly in the figure. We have

$$V_{2,a} = Z_T(j\omega_o)(I_{inj} - I_{osc,a}), \tag{2.115}$$

and

$$V_{2,b} = [Z_T(j\omega_o) + \Delta Z_T](-I_{osc,b}). \tag{2.116}$$

Note that all variables in (2.115) and (2.116) are evaluated at $\omega_{inj} = \omega_o + \Delta\omega$. Since $V_{2,a} = V_{2,b}$ while

$$I_{osc,a} = -g_m V_{2,a} \tag{2.117}$$

and

$$I_{osc,b} = -g_m V_{2,b}, \tag{2.118}$$

we have

$$I_{osc,a} = I_{osc,b}. \tag{2.119}$$

As a result, we have from (2.115) and (2.116) that

$$\frac{\Delta Z_T}{Z_T} = \frac{I_{inj}(\omega_{inj})}{I_{T,b}(\omega_{inj})}. \tag{2.120}$$

Equation (2.120) reveals that the impact of the injection signal can indeed be represented by the variation of the impedance of the resonator. Substituting (2.120) into (2.113) yields

$$|\Delta\omega| \approx \frac{\omega_o}{2Q}\left|\frac{\Delta Z_T}{Z_T}\right|. \tag{2.121}$$

Equation (2.121) shows that the lock range of the harmonic oscillator is determined by both the quality factor of the oscillator and the amount of the variation of the impedance of the resonator induced by the injection signal. For injection-locked harmonic oscillators with a spiral inductor resonator, since the value of the components of the resonator does not vary with the injection signal, the variation of the impedance caused by the injection signal is solely due to the displacement of the frequency from ω_o to $\omega_o + \Delta\omega$ in order to satisfy Barkhausen criteria. For injection-locked harmonic oscillators with an inductor inductor resonator, the inductance of the active inductors varies with the injection signal as the injection signal directly affects the transconductance of the transconductors forming the active inductors. As a result, the variation of the impedance caused by the injection signal is due to the joint effect of the displacement of the frequency $\Delta\omega$ from ω_o and the variation of the inductance of the active inductors. The latter also contributes to $\Delta\omega$.

To further investigate this important characteristic of harmonic oscillators with an active inductor resonator, let us pause for a moment and refresh our memory on active inductors [102]. Active inductors are most realized using a gyrator-C configuration that consists of a gyrator, which is a pair of cross-coupled transconductors, and a capacitor, as shown in Fig. 2.14 where g_{o1} and g_{o2} are the output conductance of transconductors 1 and 2, respectively and C_1 and C_2 are mainly the input capacitance of transconductors 1 and 2, respectively. Write Kirchhoff's Current Law at nodes 1 and 2

$$(sC_1 + g_{o1})V_1 - g_{m1}V_2 = 0, \tag{2.122}$$

$$-I_{in} + (sC_2 + g_{o2})V_2 - g_{m2}(-V_1) = 0. \tag{2.123}$$

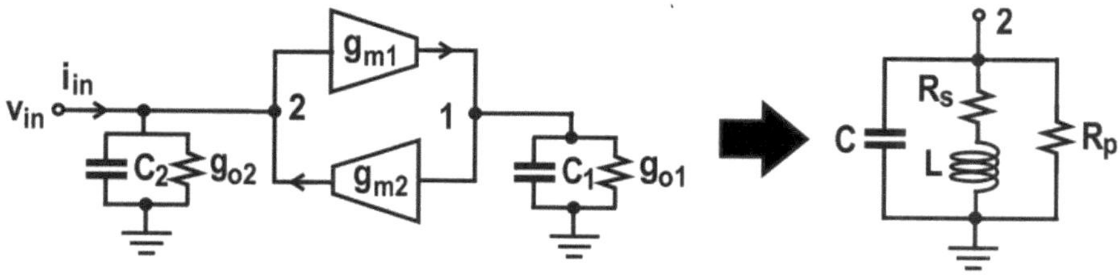

Fig. 2.14 Gyrator-C active inductors and their equivalent circuit

The admittance looking into port 2 of the gyrator-C network is obtained from

$$\begin{aligned} Y &= \frac{I_{in}}{V_{in}} \\ &= sC_2 + g_{o2} + \frac{1}{s\left(\frac{C_1}{g_{m1}g_{m2}}\right) + \frac{g_{o1}}{g_{m1}g_{m2}}}. \end{aligned} \tag{2.124}$$

Equation (2.124) can be represented by the RLC network shown in Fig. 2.14 with its components given by

$$\begin{aligned} R_p &= \frac{1}{g_{o2}}, \\ C &= C_2, \\ R_s &= \frac{g_{o1}}{g_{m1}g_{m2}}, \\ L &= \frac{C_1}{g_{m1}g_{m2}}. \end{aligned} \tag{2.125}$$

It is seen from (2.125) that the inductance of the synthesized inductor is dependent of the transconductance of the transconductors. If the output conductance of the transconductors is sufficiently small, the synthesized inductor behaves nearly lossless. Since the transconductors of active inductors are always chosen to be as simple as possible, often just a single transistor, in order to minimize power consumption and maximize operation frequency, g_{o1} and g_{o2} are typically the output resistance of MOS transistors and are therefore not small. As a result, synthesized inductors are generally lossy with a low-quality factor. This is a distinct characteristic of active inductors as compared with their spiral counterparts.

For harmonic oscillators with spiral inductors, coarse frequency tuning is typically achieved using capacitor arrays while fine frequency tuning is obtained by varying the capacitance of varactors. Since the quality factor of varactors is typically much higher than that of spiral inductors, the quality factor of these oscillators is dictated by that of the spiral inductors. For harmonic oscillators with active inductors, frequency tuning is generally obtained by varying the current of the transconductors of the gyrators constituting the active inductors subsequently their transconductance. The quality factor of these oscillators is set by that of the active inductors. Since the quality factor of active inductors is typically smaller as compared with that of spiral inductors, harmonic oscillators with active inductors generally exhibit a larger lock range as compared with those with spiral inductors. In addition to a smaller quality factor, the variation of the impedance of the resonator induced by injection signals also contributes to the lock range, as to be demonstrated using examples shortly. This unique characteristic of injection-locked harmonic oscillators with active inductors is not shared by their spiral counterparts. To further illustrate this, let the inductance of the active inductor be changed by ΔL by the injection signal. Assume that increasing the injection signal increases the current of the transconductors and subsequently their transconductance. The quality of the active inductor is estimated from

$$Q \approx \frac{\omega L}{R_s} \tag{2.126}$$

or

$$Q \approx \frac{\omega C_1}{g_{o1}}, \tag{2.127}$$

whichever dominates. It will remain unchanged when the injection signal increases. The rising of the transconductance of the transconductors lowers the inductance of the active inductor, which in turn causes the resonant frequency of the resonator to increase, as shown graphically in Fig. 2.15 where the impedance of the spiral resonator and that of the active inductor resonator are compared. The same amount of amplitude variation $|\Delta Z_T|$ and phase variation $\angle\Delta Z_T$ as those with the spiral inductor resonator need to be induced by the injection signal so as to satisfy Barkhausen criteria. The frequency displacement $\Delta\omega$ of the oscillator with the active inductor resonator consists of two parts: One due to the shift of the resonant frequency caused by the variation of the inductance of the active inductor and the other due to injection-induced amplitude and phase variations. Note that the shift of the resonant frequency caused by the variation of the inductance of the active inductor does not change the amplitude and phase of the impedance of the resonator.

To validate the preceding results, consider two oscillators, one with a spiral inductor resonator and the other with an active inductor resonator [112]. Figure 2.16 shows the simplified schematic of the oscillators. Both oscillators are designed in an IBM 130 nm 1.2 V CMOS technology. The injection signal of both oscillators is a sinusoid of amplitude −18 dBm and frequency 2.4 GHz. The negative resistors of both oscillators have the same transistor dimension and DC biasing current. The swing of the output voltage of the oscillators is also made the same. Figure 2.17 shows the dependence of the lock range of the oscillators on the strength of the injection signal. It is observed that the lock range of both oscillators increases with rising injection strength linearly in the range of investigation. Also observed is that the lock range of the oscillator with an active-inductor resonator is significantly larger as compared with that with a spiral inductor resonator. Further observed is that the ratio of the lock range of the oscillator with the active inductor resonator is

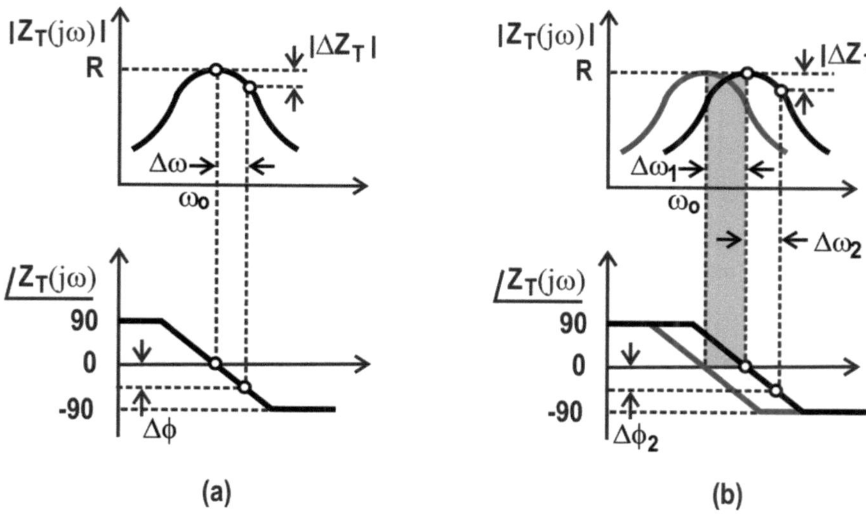

Fig. 2.15 (**a**) Impedance of spiral inductor resonator. $\Delta\omega$ is the injection-induced frequency shift in order to satisfy Barkhausen criteria. (**b**) Impedance of active inductor resonator. $\Delta\omega_1$ is the shift of the resonant frequency of the resonator due to the injection-induced variation of the impedance of the resonator. $\Delta\omega_2$ and $\Delta\phi_2$ are the injection-induced frequency shift and phase shift, respectively in order to satisfy Barkhausen criteria. Note $\Delta\omega$ in (**a**) is the same as $\Delta\omega_2$ in (**b**). Similarly, $\Delta\phi$ in (**a**) is the same as $\Delta\phi_2$ in (**b**). Also note that in (**b**) we assume $|Z_T(j\omega_o)| = |Z_T[j(\omega_o + \Delta\omega_1)]$

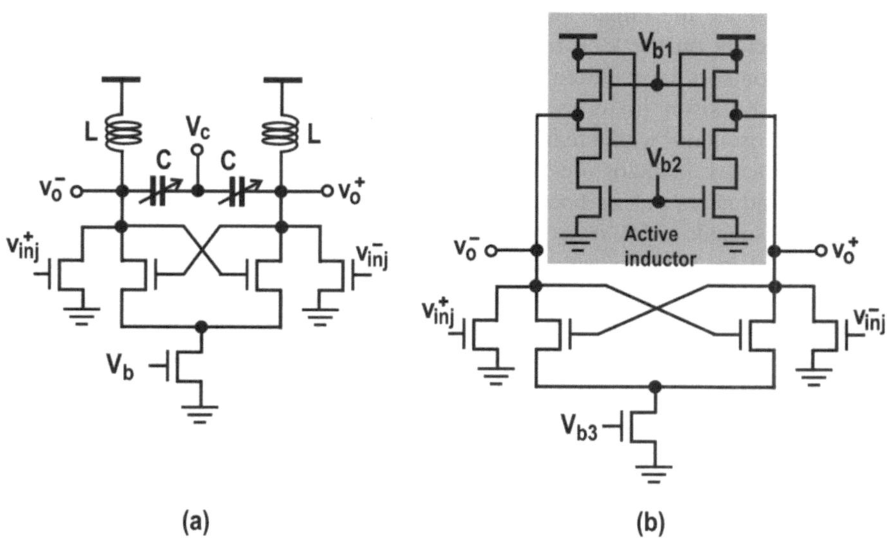

Fig. 2.16 Left (**a**): Harmonic oscillator with spiral inductor resonators. Right (**b**): Harmonic oscillator with active inductor resonators

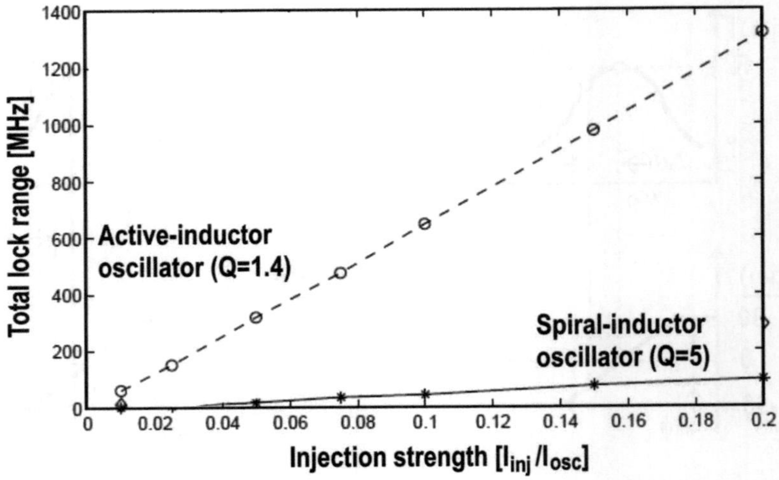

Fig. 2.17 Comparison of the lock range of harmonic oscillators with spiral inductor and active inductor resonators

over 10 times that with the spiral inductor resonator even though the quality factor of the active inductor is 28% of that of the spiral inductor, revealing that the large lock range of the oscillator with the active inductor resonator is not just caused by its small quality factor but also due to the injection-induced inductance variation of the active inductor resonator.

To investigate the impact of the injection-induced impedance variation of the active inductors on the lock range, the quality factor of the oscillator with the passive resonators is made comparable to that of the active inductor oscillator by adding a resistor in series with the spiral inductors. The lock range of the passive LC oscillator with a reduced quality factor is shown in Fig. 2.18, along with that of the active inductor oscillator. It is seen that lowering the quality factor of the spiral inductors indeed boosts the lock range of the oscillator with the spiral inductor resonator. Also observed is that the lock range of the oscillator with low-Q spiral inductor resonators is still much smaller as compared with that of the active-inductor oscillator despite of their comparable quality factor especially when injection strength is high, confirming that the lock range of the harmonic oscillator with an active inductor resonator is indeed set by both the quality factor of the resonators and the variation of the impedance of the resonators induced by the injection signal. The larger the injection signal, the larger the variation of the impedance of the resonators, the larger the lock range.

The large lock range of the oscillator with active inductor resonators is attractive when these oscillators are used as the pre-scalar of frequency synthesizers if the additional amount of the power consumption of the active inductors is acceptable. One might argue that as active inductors contain a large number of active devices, the noise performance of oscillators with active inductor resonators is inferior as compared with oscillators with spiral inductor resonators. Fortunately the noise

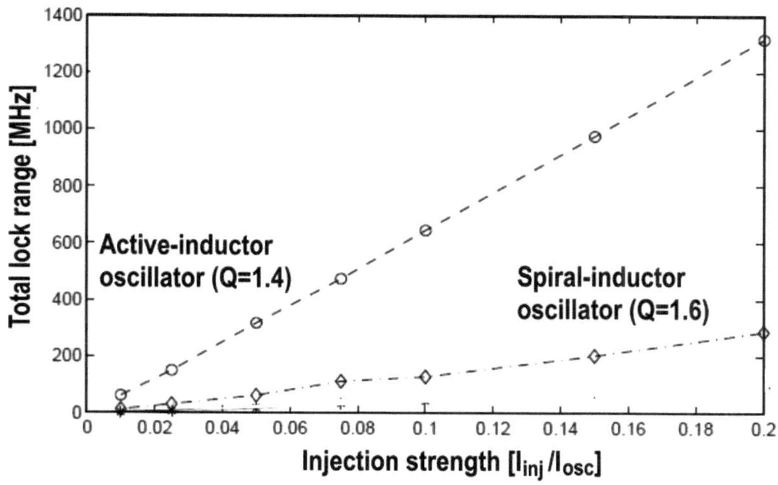

Fig. 2.18 Impact of the injection-induced impedance variation of the active inductor resonator on the lock range of harmonic oscillators with the active inductor resonator

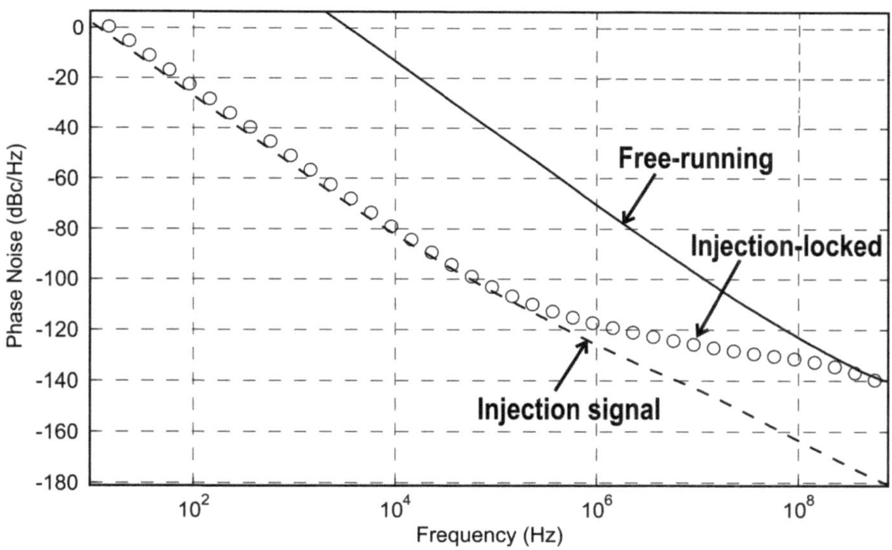

Fig. 2.19 Simulated phase noise of harmonic oscillator with active-inductor resonators

performance of an injection-locked harmonic oscillator is dictated by that of the injection signal rather than that of the oscillator, attributed to the existence of a large bandwidth feedback mechanism in the injection-locked oscillator that corrects any deviation of the frequency of the oscillator from that of the injection signal [64]. To demonstrate this, Fig. 2.19 plots the phase noise of the harmonic oscillator with an active-inductor resonator in Fig. 2.16. It is seen that the phase noise of the oscillator in the free-running state is significantly higher as compared with that of the injection signal. When the oscillator is locked to the injection signal, its phase noise will become comparable to that of the injection signal if the difference between the frequency of the injection signal and the free-running frequency of the oscillator is small. Their phase noise difference will become large when the frequency of the injection signal deviates from the free-running frequency of the oscillator. When the frequency difference becomes sufficiently large, the phase noise of the injection-locked oscillator will approach that of the oscillator in the free-running state.

2.6 Superharmonic Injection

In the preceding analysis, the frequency of injection signals is in the vicinity of the natural frequency of harmonic oscillators under injection. The frequency of the injection signal of a harmonic oscillator can also be a harmonic of the free-running frequency of the oscillator, i.e., $N\omega_o$ where ω_o is the free-running frequency of the oscillator and $N = 2, 3, \ldots$ is an integer. This is known as superharmonic injection. The frequency of the injection signal of a harmonic oscillator can also be a fraction of the free-running frequency of the oscillator, i.e., ω_o/N. This is known as subharmonic injection. For a superharmonic injection-locked harmonic oscillator, since the frequency of the injection signal is a multiple of that of the oscillator, the oscillator functions as an integer frequency divider. Similarly for a subharmonic injection-locked harmonic oscillator, since the frequency of the injection signal is a fraction of that of the oscillator, the oscillator functions as a frequency multiplier. This section is concerned with the superharmonic injection-locking of harmonic oscillators. The subharmonic injection-locking of harmonic oscillators will be dealt with in the next section.

Perhaps the most successful application of superharmonic injection-locking is injection-locked frequency dividers (ILFD). In an integer-N frequency synthesizer, a frequency divider with a large division ratio is needed in the feedback path to generate a low-frequency signal from the output of the high-frequency oscillator of the synthesizer so that the frequency of the generated low-frequency signal can be compared with that of a low-frequency reference, often from the output of a crystal oscillator, such that the frequency of the oscillator can be stabilized via negative feedback. A large frequency division ratio is often obtained using a multistage approach in which multiple frequency dividers are cascaded. Among the stages of the frequency dividers, the first stage is most difficult to design and also consumes the most power. This is because not only the first-stage must operate at the same frequency as that of the oscillator of the synthesizer, it must also cope with a large variation of the frequency of its input arising from the impact of PVT (process, voltage, and temperature) uncertainty on the frequency of the oscillator of the synthesizer. It becomes increasingly difficult to design frequency dividers operating at high frequencies using conventional digital logic such as current-mode logic. Attributive to the intrinsic characteristics of ILFDs including high operation frequency, low power consumption, and excellent phase noise, ILFDs are widely popular in realization of the first stage of the divide-by-N block of frequency synthesizers, which is known as the pre-scalar.

Harmonic oscillators are favored over their nonharmonic counterparts in implementing ILFDs for wireless applications due to their high frequency selectivity subsequently better phase noise performance. The high frequency selectivity of harmonic oscillators, owing to the high-quality factor of their LC resonators, also ensures that the output frequency of the frequency divider is the desired frequency, which is the center frequency of the passband of the LC resonators. Unlike harmonic oscillators, the low-quality factor of nonharmonic oscillators such as ring oscillators or relaxation oscillators permits the existence of a large number of frequency components in the oscillators. When mixed with the injection signal, the lack of an effective frequency selection mechanism in these oscillators will give rise to a high level of uncertainty in the output frequency of the oscillators. As a result, these oscillators are prone to locking to a wrong harmonic subsequently an incorrect frequency division ratio. The phase noise performance of nonharmonic oscillators is also inferior as compared with that of harmonic oscillators under the same injection strength, further making them less attractive for wireless applications.

In this section, we investigate second-order and third-order superharmonic injection-locking of harmonic oscillators from a theoretical perspective, similar to what we did for first-harmonic injection-locking. Our attention will then turn to the design of divide-by-2 and divide-by-3 injection-locked frequency dividers in the late chapters.

2.6.1 Second-Order Superharmonic Injection

Let us first consider that the frequency of the injection signal is approximately twice that of the free-running frequency of the oscillator under injection

$$
\begin{aligned}
i_{inj}(t) &= I_{inj} \cos\left(2\omega t + \phi_{inj}\right) \\
&= \frac{I_{inj}}{2} e^{j2\omega t} e^{j\phi_{inj}} + \frac{I_{inj}}{2} e^{-j2\omega t} e^{-j\phi_{inj}}.
\end{aligned} \tag{2.128}
$$

with ω in the vicinity of the free-running frequency of the oscillator. Again, the block diagram of the oscillator is shown in Fig. 2.9 with the characteristics of the nonlinear block depicted by (2.66). When the oscillator is locked to the injection signal, its frequency will be ω. The output of the oscillator in the lock state can be written as

$$\begin{aligned} i_o &= I_o \cos(\omega t) \\ &= \frac{I_o}{2} e^{j\omega t} + \frac{I_o}{2} e^{-j\omega t}. \end{aligned} \tag{2.129}$$

The input of the nonlinear block of the oscillator with the presence of the injection signal is given by

$$\begin{aligned} x &= i_{inj} - i_o \\ &= \frac{I_{inj}}{2} e^{j2\omega t} e^{j\phi_{inj}} + \frac{I_{inj}}{2} e^{-j2\omega t} e^{-j\phi_{inj}} - \frac{I_o}{2} e^{j\omega t} - \frac{I_o}{2} e^{-j\omega t}. \end{aligned} \tag{2.130}$$

The output of the nonlinear block is given by

$$\begin{aligned} y = a_1 &\left(\frac{I_{inj}}{2} e^{j2\omega t} e^{j\phi_{inj}} + \frac{I_{inj}}{2} e^{-j2\omega t} e^{-j\phi_{inj}} - \frac{I_o}{2} e^{j\omega t} - \frac{I_o}{2} e^{-j\omega t} \right) \\ + a_2 &\left(\frac{I_{inj}}{2} e^{j2\omega t} e^{j\phi_{inj}} + \frac{I_{inj}}{2} e^{-j2\omega t} e^{-j\phi_{inj}} - \frac{I_o}{2} e^{j\omega t} - \frac{I_o}{2} e^{-j\omega t} \right)^2 \\ + a_3 &\left(\frac{I_{inj}}{2} e^{j2\omega t} e^{j\phi_{inj}} + \frac{I_{inj}}{2} e^{-j2\omega t} e^{-j\phi_{inj}} - \frac{I_o}{2} e^{j\omega t} - \frac{I_o}{2} e^{-j\omega t} \right)^3. \end{aligned} \tag{2.131}$$

Among the frequency components of the output of the nonlinear block, only those that fall into the passband of the resonator pass through the resonator and reach the output of the oscillator. These frequency components are

- $-\frac{a_1 I_o}{2} e^{j\omega t}$,
- $-2a_2 \left(\frac{I_{inj}}{2}\right)\left(-\frac{I_o}{2}\right) e^{j\phi_{inj}} e^{j\omega t}$, and
- $-3a_3 \left(-\frac{I_o}{2}\right)^2 \left(-\frac{I_o}{2}\right) e^{j\omega t}$.

All other frequency components of the output of the nonlinear block are suppressed by the resonator. We therefore have the following phasor-domain relation at ω

$$\left(-\frac{a_1}{2} I_o - \frac{a_2}{2} I_{inj} I_o e^{j\phi_{inj}} - \frac{3a_3}{8} I_o^3 \right) \frac{H_o}{1 + j2Q\left(\frac{\omega - \omega_o}{\omega_o}\right)} = \frac{I_o}{2}. \tag{2.132}$$

Equating the real part of (2.132) yields

$$1 + a_1 H_o + a_2 H_o I_{inj} \cos\phi_{inj} + \frac{3a_3}{4} H_o I_o^2 = 0. \tag{2.133}$$

Equation (2.133) is Barkhausen magnitude criterion. To derive lock range

$$\Delta\omega_{L,2nd,super} = \omega - \omega_o, \tag{2.134}$$

we equate the imaginary part of (2.132)

$$\Delta\omega_{L,2nd,super} = -\frac{\omega_o}{2Q}(a_2 H_o) I_{inj} \sin\phi_{inj}. \tag{2.135}$$

It is seen from (2.135) that the larger the injection signal, the larger the lock range. Also, the larger the second-order nonlinearity signified by a_2, the larger the lock range. Moreover, the larger the gain of the resonator, the larger the lock range. The reason for this is that if the gain of the resonator is large, since the loop gain needs to be unity in order to for the oscillator to oscillate at ω, a large frequency displacement from the center frequency of the resonator ω_o is needed in order to generate a sufficiently large amplitude drop in the impedance of the resonator so as to satisfy Barkhausen magnitude criterion.

The lock range is also a function of the injection phase. The lock range will be maximized when

$$\phi_{inj}^* = \frac{\pi}{2}. \tag{2.136}$$

The maximum lock range is given by

$$\Delta\omega_{L,2nd,super,max} \approx -\frac{\omega_o}{2Q}(a_2 H_o) I_{inj}. \tag{2.137}$$

Figure 2.20 plots i_{inj} and i_o with $\phi_{inj} = \pi/2$. ϕ_{inj} is measured with reference to 2ω. It is seen that at $\phi_{inj} = \pi/2$, i_{inj} will peak when i_o crosses the threshold, thereby asserting the maximum impact on the frequency of the oscillator, resulting in the maximum lock range.

At the optimal injection phase, Eq. (2.133) becomes

$$1 + a_1 H_o + \frac{3a_3}{4} H_o I_o^2 = 0. \tag{2.138}$$

Equation (2.148) is the same as that in first-harmonic injection-locking given by (2.98).

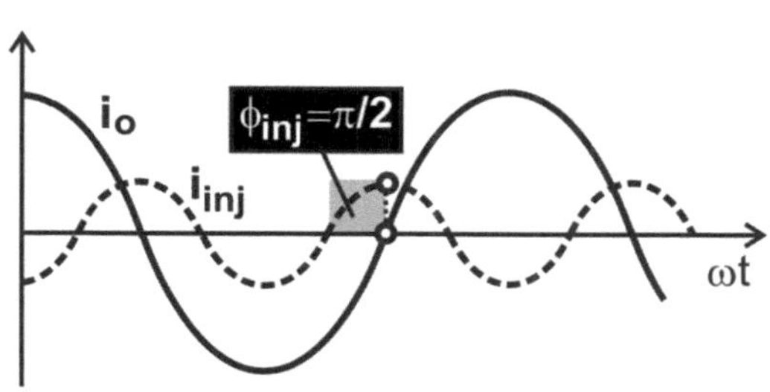

Fig. 2.20 Second-order superharmonic injection with i_{inj} leading i_o by $\pi/2$

2.6.2 Third-Order Superharmonic Injection

Now let us consider that the frequency of the injection signal is approximately three times that of the free-running frequency of the oscillator under injection

$$\begin{aligned} i_{inj}(t) &= I_{inj}\cos\left(3\omega t+\phi_{inj}\right) \\ &= \frac{I_{inj}}{2}e^{j3\omega t}e^{j\phi_{inj}}+\frac{I_{inj}}{2}e^{-j3\omega t}e^{-j\phi_{inj}}. \end{aligned} \tag{2.139}$$

with ω in the vicinity of ω_o. The output of the oscillator in the lock state is given by

$$\begin{aligned} i_o &= I_o\cos(\omega t) \\ &= \frac{I_o}{2}e^{j\omega t}+\frac{I_o}{2}e^{-j\omega t}. \end{aligned} \tag{2.140}$$

The block diagram of the oscillator is shown in Fig. 2.9 with the characteristics of the nonlinear block depicted by (2.66). The input of the nonlinear block of the oscillator with the presence of the injection signal is given by

$$\begin{aligned} x &= i_{inj}-i_o \\ &= \frac{I_{inj}}{2}e^{j3\omega t}e^{j\phi_{inj}}+\frac{I_{inj}}{2}e^{-j3\omega t}e^{-j\phi_{inj}}-\frac{I_o}{2}e^{j\omega t}-\frac{I_o}{2}e^{-j\omega t}. \end{aligned} \tag{2.141}$$

The output of the nonlinear block is obtained from

$$\begin{aligned} y = {} & a_1\left(\frac{I_{inj}}{2}e^{j3\omega t}e^{j\phi_{inj}}+\frac{I_{inj}}{2}e^{-j3\omega t}e^{-j\phi_{inj}}-\frac{I_o}{2}e^{j\omega t}-\frac{I_o}{2}e^{-j\omega t}\right) \\ & +a_2\left(\frac{I_{inj}}{2}e^{j3\omega t}e^{j\phi_{inj}}+\frac{I_{inj}}{2}e^{-j3\omega t}e^{-j\phi_{inj}}-\frac{I_o}{2}e^{j\omega t}-\frac{I_o}{2}e^{-j\omega t}\right)^2 \\ & +a_3\left(\frac{I_{inj}}{2}e^{j3\omega t}e^{j\phi_{inj}}+\frac{I_{inj}}{2}e^{-j3\omega t}e^{-j\phi_{inj}}-\frac{I_o}{2}e^{j\omega t}-\frac{I_o}{2}e^{-j\omega t}\right)^3. \end{aligned} \tag{2.142}$$

The frequency components of the output of the nonlinear block that fall into the passband of the resonator are:

- $-a_1\frac{I_o}{2}e^{j\omega t}$,
- $3a_3\left(-\frac{I_o}{2}\right)^2\left(\frac{I_{inj}}{2}\right)e^{j\phi_{inj}}e^{j\omega t}$, and
- $3a_3\left(-\frac{I_o}{2}\right)^2\left(-\frac{I_o}{2}\right)e^{j\omega t}$.

We therefore have the following phasor-domain relation at ω

$$\left(-\frac{a_1}{2}I_o+\frac{3a_3}{8}I_{inj}I_o^2e^{j\phi_{inj}}-\frac{3a_3}{8}I_o^3\right)\frac{H_o}{1+j2Q\left(\frac{\omega-\omega_o}{\omega_o}\right)}=\frac{I_o}{2}. \quad (2.143)$$

Equating the real part of (2.143) yields

$$1+a_1H_o+\frac{3a_3}{4}H_oI_o^2-\frac{3a_3}{4}H_oI_{inj}I_o\cos\phi_{inj}=0. \quad (2.144)$$

Similarly, equating the imaginary part of (2.143) gives the lock range

$$\Delta\omega_{L,3rd,super}=\frac{\omega_o}{2Q}\left(\frac{3a_3}{4}H_o\right)I_{inj}I_o\sin\phi_{inj}. \quad (2.145)$$

It is seen from (2.145) that the larger the injection signal, the larger the lock range. Also, the larger the third-order nonlinearity quantified by a_3, the larger the lock range. Moreover, the larger the output of the oscillator, the larger the lock range. It is also seen that the larger the gain of the resonator, the larger the lock range. The reason for this was provided in the second-order superharmonic injection of harmonic oscillators presented earlier.

The lock range is also injection-phase dependent. It will be maximized when

$$\phi_{inj}^*=\frac{\pi}{2}. \quad (2.146)$$

The maximum lock range is given by

$$\Delta\omega_{L,3rd,super,max}=\frac{\omega_o}{2Q}\left(\frac{3a_3}{4}H_o\right)I_{inj}I_o. \quad (2.147)$$

Figure 2.21 shows the waveforms of i_{inj} and i_o with $\phi_{inj}=\pi/2$. Note that ϕ_{inj} is measured with reference to 3ω. It is seen that at $\phi_{inj}=\pi/2$, i_{inj} peaks while i_o crosses the threshold. i_{inj} therefore asserts the maximum impact on the frequency of the oscillator. This is echoed with the maximum lock range.

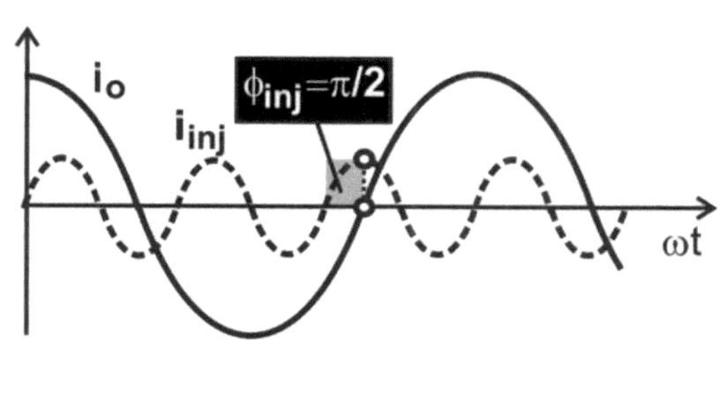

Fig. 2.21 Third-order superharmonic injection with i_{inj} leading i_o by $\pi/2$

At the optimal injection phase, Eq. (2.144) becomes

$$1 + a_1 H_o + \frac{3a_3}{4} H_o I_o^2 = 0. \tag{2.148}$$

Equation (2.148) is the same as that in second-order superharmonic injection given by (2.138) and that in first-harmonic injection given by (2.98).

2.6.3 *Divide-by-2 Injection-Locked Frequency Dividers*

A divide-by-2 ILFD has its input at frequency 2ω and output at frequency ω. Figure 2.22 shows the block diagram of the divide-by-2 ILFD. Note the different representations of the frequency division mechanism that generates the desired frequency component in Fig. 2.22 and the block diagram of injection-locked harmonic oscillators in Fig. 2.9. The summer and nonlinear block in Fig. 2.9 are now replaced with the mixer in Fig. 2.22 to emphasize explicitly that it is the mixing operation that generates the desired frequency component. It should, however, be noted that frequency mixing is generally done utilizing nonlinear characteristics of MOS devices rather than employing an explicit frequency mixer. Figures 2.9 and 2.22 convey essentially the same information except that Fig. 2.22 is more explicit.

The ILFD consists of a mixer that mixes the input signal of the frequency divider at ω_{inj} and the output of the frequency divider at $\omega_{inj}/2$, and a high-Q band-pass filter whose passband center frequency is set to the desirable output frequency of the frequency divider, i.e., $\omega_{inj}/2$. At the output of the mixer, frequency components at $\omega_{inj}/2$ and $3\omega_{inj}/2$ exist if the mixer is perfectly linear. When the nonlinear characteristics of the mixer are accounted for, frequency components other than $\omega_{inj}/2$ and $3\omega_{inj}/2$ will also exist at the output of the mixer. The band-pass filter has a sufficiently large quality factor such that it filters out all unwanted frequency components generated in mixing operation. Only frequency components $\omega_{inj}/2$ pass through the filter without attenuation and reach the output of the frequency divider, thereby achieving a frequency division ratio of 2.

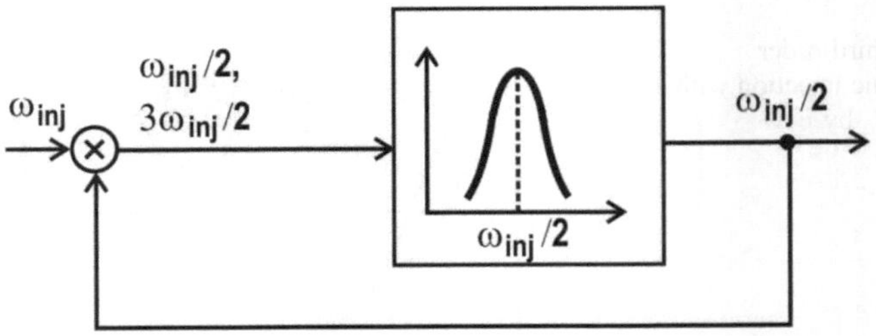

Fig. 2.22 Block diagram of divide-by-2 ILFDs

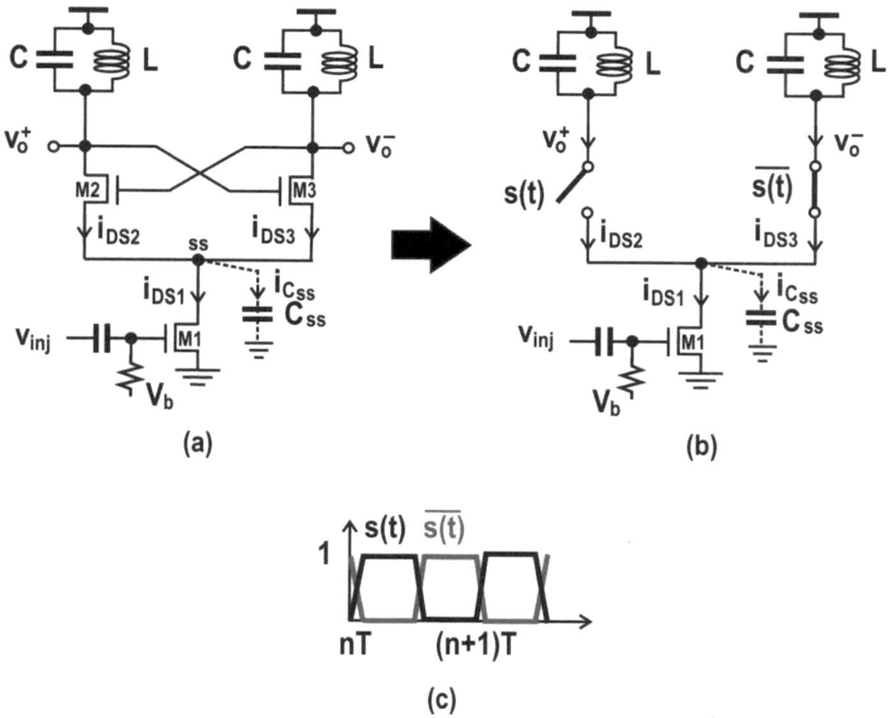

Fig. 2.23 (**a**) Divide-by-2 ILFD. (**b**) M2 and M3 are assumed to operate in on ON/OFF mode such that they are modeled as ideal switches clocked by switching functions $s(t)$ and $\overline{s(t)}$. (**c**) Waveforms of switching functions $s(t)$ and $\overline{s(t)}$

Although various architectures available to implement mixers, both explicit or implicit, in order to minimize power consumption and at the same time to cope with stringent constraints imposed by the high frequency of the input when ILFDs are used as the pre-scalar of frequency synthesizers, mixers in injection-locked frequency dividers are typically realized using a single transistor and mixing is achieved utilizing the nonlinear characteristics of the transistor. Both gate-pumped and drain-pumped architectures are used. The former utilizes the nonlinear characteristics of $i_{DS} \sim v_{GS}$ relation of the transistor when v_{GS} is in the vicinity of the threshold voltage of the transistor, whereas the latter takes the advantage of the nonlinear relation of $i_{DS} \sim v_{DS}$ in the vicinity of the pinch-off (knee) mode of the transistor [19]. The band-pass filter is typically realized using a LC resonator, compensated with a synthesized negative resistor. The ILFD is a second-order superharmonic injection-locked harmonic oscillator.

Figure 2.23a shows the schematic of a divide-by-2 injection-locked frequency divider. The input of the frequency divider is the injection signal applied to the gate of the tail current source transistor M1 biased in saturation by V_b. The value of V_b

and the dimension of M1 are set as per the desired value of the negative resistance needed to compensate for the resistive loss of the LC resonators. Assume that the frequency of the output of the frequency divider is ω. We have

$$\begin{aligned} v_{o+} &= V_{o,DC} + V_o \cos\omega t, \\ v_{o-} &= V_{o,DC} - V_o \cos\omega t, \end{aligned} \tag{2.149}$$

where $V_{o,DC}$ is the common-mode voltage of the output of the oscillator and V_o is the amplitude of the AC output voltage of the oscillator. Since

$$\begin{aligned} i_{DS2} &= k_{n2}\left(V_{o,DC} - V_o\cos\omega t - V_{ss} - V_T\right)^2 \\ &= k_{n1}\left(V_{o,DC} - V_{ss} - V_T\right)^2\left[1 - \frac{2V_o\cos\omega t}{V_{o,DC} - V_{ss} - V_T} + \frac{V_o^2\cos^2\omega t}{(V_{o,DC} - V_{ss} - V_T)^2}\right] \end{aligned} \tag{2.150}$$

and

$$\begin{aligned} i_{DS3} &= k_{n3}\left(V_{o,DC} + V_o\cos\omega t - V_{ss} - V_T\right)^2 \\ &= k_{n2}\left(V_{o,DC} - V_{ss} - V_T\right)^2\left[1 + \frac{2V_o\cos\omega t}{V_{o,DC} - V_{ss} - V_T} + \frac{V_o^2\cos^2\omega t}{(V_{o,DC} - V_{ss} - V_T)^2}\right], \end{aligned} \tag{2.151}$$

where

$$k_{n2,3} = \frac{1}{2}\mu_n C_{ox}\left(\frac{W}{L}\right)_{2,3}, \tag{2.152}$$

and V_{ss} is the voltage at the sources of M2 and M3, we have

$$\begin{aligned} i_{DS1} &= i_{DS2} + i_{DS3} \\ &= 2k_n\left(V_{o,DC} - V_{ss} - V_T\right)^2\left[1 + \frac{V_o^2\cos^2\omega t}{(V_{o,DC} - V_{ss} - V_T)^2}\right] \\ &= 2k_n\left(V_{o,DC} - V_{ss} - V_T\right)^2\left\{1 + \frac{V_o^2}{2(V_{o,DC} - V_{ss} - V_T)^2}\left[1 + \cos(2\omega t)\right]\right\}. \end{aligned} \tag{2.153}$$

Note that we have assumed $k_{n2} = k_{n3} = k_n$, i.e., transistors M2 and M3 are perfectly matched. It is seen from (2.153) that i_{DS1} only contains a DC component and the second harmonic of the frequency of the oscillator. This observation suggests if v_{inj} is a single tone at 2ω, the tail current will be at 2ω as well, but the output frequency of the oscillator will be at ω.

To further illustrate this, we assume that the voltage swing of the output of the oscillator is sufficiently large such that transistors M2 and M3 operate in an ON/OFF mode. They can therefore be modeled as ideal switches to simplify analysis. The operation of M2 and M3 is depicted using switching function $s(t)$ and its complementary $\overline{s(t)}$. Both are periodic with period T and a 50% duty cycle, as shown graphically in Fig. 2.23c. Note that the symmetry of the cross-coupled transistors and the fact that the common-source node of M2 and M3 is an AC ground ensure that the duty cycle of the output of the oscillator is 50%. Since $s(t)$ and $\overline{s(t)}$ are periodic, they can be represented by Fourier series

$$s(t) = \frac{1}{2} + \frac{2}{\pi} \sum_{n=1,3,5,...}^{\infty} \frac{1}{n} \sin(n\omega)t, \tag{2.154}$$

and

$$\overline{s(t)} = \frac{1}{2} - \frac{2}{\pi} \sum_{n=1,3,5,...}^{\infty} \frac{1}{n} \sin(n\omega)t, \tag{2.155}$$

where $\omega = \frac{2\pi}{T}$. It is seen that $s(t)$ and $\overline{s(t)}$ only contain the odd harmonics of ω.

Let the injection signal be a sinusoid of frequency 2ω

$$v_{inj}(t) = V_{inj} \cos(2\omega t). \tag{2.156}$$

The current of M1 is obtained from

$$\begin{aligned} i_{DS1} &= k_{n1} \left[V_{inj} \cos(2\omega t) + V_b - V_T\right]^2 \\ &= k_{n1} (V_b - V_T)^2 \left[1 + \frac{2V_{inj} \cos(2\omega t)}{V_b - V_T} + \left(\frac{V_{inj}}{V_b - V_T}\right)^2 \cos^2(2\omega t)\right]. \end{aligned} \tag{2.157}$$

i_{DS1} is completely steered between M2 and M3 by $s(t)$ and $\overline{s(t)}$. If i_{DS1} is not completely steered between M2 and M3 by $s(t)$ and $\overline{s(t)}$, i.e., M2 and M3 do not operate in an ON/OFF mode, they will conduct both a DC current and an AC current. The AC current is still completely steered between M2 and M3. The DC current does not contribute to frequency mixing subsequently frequency division. The modulated currents, which are the channel currents of M2 and M3, are given by

$$\begin{aligned} i_{DS2} &= s(t) i_{DS1} \\ &= k_{n1} (V_b - V_T)^2 \left[1 + \frac{2V_{inj} \cos(2\omega t)}{V_b - V_T} + \left(\frac{V_{inj}}{V_b - V_T}\right)^2 cos^2(2\omega t)\right] \end{aligned}$$

$$\times\left[\frac{1}{2}+\frac{2}{\pi}\sum_{n=1,3,5,\ldots}^{\infty}\frac{1}{n}\sin(n\omega)t\right]$$

$$\approx k_{n1}\left(V_b-V_T\right)^2\left[1+\frac{2V_{inj}\cos(2\omega t)}{V_b-V_T}\right]\left[\frac{1}{2}+\frac{2}{\pi}\sum_{n=1,3,5,\ldots}^{\infty}\frac{1}{n}\sin(n\omega)t\right], \tag{2.158}$$

and

$$\begin{aligned} i_{DS3} &= \overline{s(t)}i_{DS1} \\ &= k_{n1}\left(V_b-V_T\right)^2\left[1+\frac{2V_{inj}\cos(2\omega t)}{V_b-V_T}+\left(\frac{V_{inj}}{V_b-V_T}\right)^2 cos^2(2\omega t)\right] \\ &\quad\times\left[\frac{1}{2}-\frac{2}{\pi}\sum_{n=1,3,5,\ldots}^{\infty}\frac{1}{n}\sin(n\omega)t\right]. \\ &\approx k_{n1}\left(V_b-V_T\right)^2\left[1+\frac{2V_{inj}\cos(2\omega t)}{V_b-V_T}\right]\left[\frac{1}{2}-\frac{2}{\pi}\sum_{n=1,3,5,\ldots}^{\infty}\frac{1}{n}\sin(n\omega)t\right]. \end{aligned} \tag{2.159}$$

Note that we have neglected the second-order term in simplification of (2.158) and (2.159). Since i_{DS2} and i_{DS3} flow to the LC resonators whose self-resonant frequency is ω, the resonators suppress all frequency components except those at ω. As a result, a frequency division ratio of 2 is obtained.

Table 2.1 shows the frequency components of the injection signal and those of the switching signals. It is seen that it is the mixing of the fundamental and third harmonic components of the switching signals with the injection signal at 2ω that generates the desired frequency component at ω. Because the amplitude of the harmonic components of the switching signals decreases with the increase in the order of the harmonics, it is the mixing of the fundamental component of the switching signals with the injection signal at 2ω that constitutes most of the desired frequency component at ω. Frequency components that are crossed out in the table fall outside the passband of the LC resonators and are therefore filtered out by the LC resonators.

The preceding divide-by-2 ILFD is not attractive for high-frequency applications. This is because a large capacitance C_{ss} is encountered at the common-source node SS of M2 and M3, arising from the capacitances of M1, M2, and M3. This large capacitance shunts a large portion of the high-frequency injection-induced channel current of M1 to the ground, thereby lowering the amount of the current flowing to M2 and M3 that actually mixes with the switching signal, thereby lowering injection

Table 2.1 Frequency components of divide-by-2 ILFD

Injection signal	Switching signal	Frequency addition ($\omega_{inj} + \omega_s$)	Frequency subtraction ($\lvert\omega_{inj} - \omega_s\rvert$)
2ω	ω	~~3ω~~	ω
2ω	3ω	~~5ω~~	ω
2ω	5ω	~~7ω~~	~~3ω~~
2ω	7ω	~~9ω~~	~~5ω~~
...	...	...	...

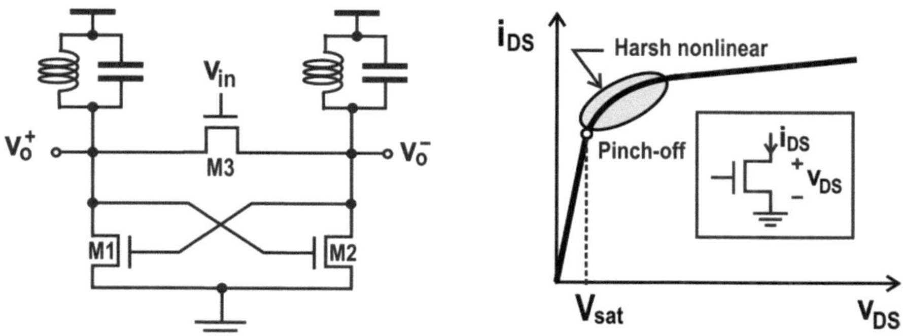

Fig. 2.24 Divide-by-2 ILFD using a drain-pumped configuration

efficiency subsequently lock range. This frequency divider is also not attractive for low-voltage applications due to the large drain-source voltage drop of M1 required to ensure that M1 operates in saturation.

To eliminate the aforementioned drawbacks of the ILFD with injection signal applied to the gate of the tail current source transistor, an injection signal can be injected directly into the output nodes of the resonators, as shown in Fig. 2.24 [8, 14, 18, 45, 92, 94, 96]. Since the injection signal is directly injected into the output nodes of the LC resonators without going through circuitry that might attenuate the injection signal prior to the mixing operation where the desired frequency component is generated, injection efficiency is improved. The improved injection efficiency also permits the use of small injection transistor M3. This in turn minimizes the detrimental effect of the capacitances of the injection transistor, making this topology particularly attractive for high-frequency applications. Note that the dimensions of M1 and M2 are dictated by the required negative resistance to compensate for the resistive loss of the resonator. They are therefore fixed for a given resonator. The elimination of the tail current source transistor also eliminates the up-conversion of the flicker noise of the transistor, thereby improving the phase noise performance of the ILFD.

If the injection signal is sufficiently large, M3 will operate in triode mode, forcing $v_{o+} \approx v_{o-}$ and in cut-off mode when the injection signal is low. As a result, the

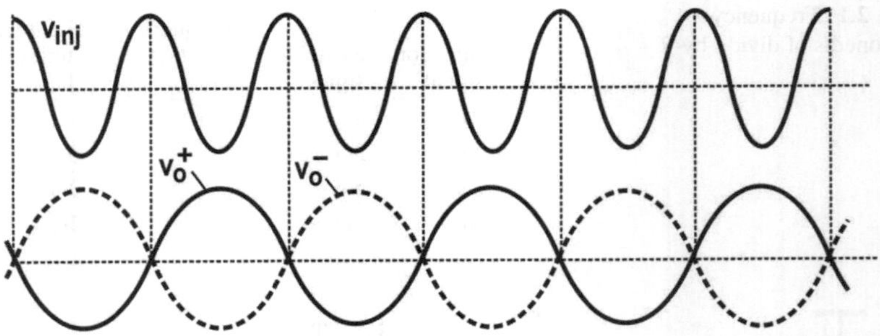

Fig. 2.25 Divide-by-2 ILFD using a drain-pumped configuration with a strong injection signal

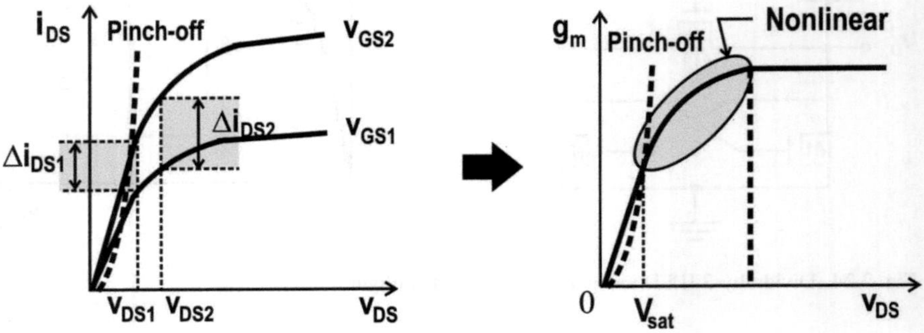

Fig. 2.26 Nonlinear transconductance of MOS transistors

injection signal equalizes the outputs each time v_{inj} peaks, a division ratio of 2 is therefore obtained, as shown graphically in Fig. 2.25 [96].

If the injection signal is not sufficiently large, the injection transistor will not operate in an ON/OFF mode but rather in saturation mode. Mixing operation in this case is based on the principle of drain-pumped mixing [7, 10]. Drain-pumped mixing utilizes the nonlinear characteristics of $i_{DS} \sim v_{DS}$ of MOS transistors in saturation with their operation point just outside the pinch-off point of the transistors, as illustrated graphically in Fig. 2.24. The large variation of the voltage of v_{o+} and v_{o-} subsequently the large variation of v_{DS3} allows M3 to function as a drain-pumped mixer that mixes v_{in} with $v_o = v_{DS3} = v_o^+ - v_o^-$. The main nonlinearity of the drain-pumped mixer is g_m, which is modulated by v_{DS}, as shown graphically in Fig. 2.26. Because

$$g_m = \frac{\partial i_{DS}}{\partial v_{GS}} \approx \frac{\Delta i_{DS}}{\Delta v_{GS}}, \tag{2.160}$$

for the same $\Delta v_{GS} = v_{GS2} - v_{GS1}$, Δi_{DS} varies with v_{DS}, as illustrated in Fig. 2.26. It is seen that the larger v_{DS}, the larger Δi_{DS} subsequently the larger g_m.

In the region where the transistor is in triode, since i_{DS} varies with v_{DS} linearly, for the same Δv_{GS}, the transconductance at given v_{DS} is obtained from

$$g_m \approx \frac{\Delta i_{DS}}{\Delta v_{GS}} = \left(\frac{g_{ds2} - g_{ds1}}{\Delta v_{GS}}\right) v_{DS}, \tag{2.161}$$

where g_{ds2} and g_{ds1} are the slope of the $i_{DS} \sim v_{DS}$ curve with the gate-source voltage at v_{GS2} and v_{GS1}, respectively. Equation (2.161) shows that the relation between g_m and v_{DS} is linear when the device is in triode.

In the region where the transistor is saturation but in the vicinity of the pinch off, the relation between i_{DS} and v_{DS} is nonlinear and is depicted by $i_{DS} = f(v_{DS})$. Since

$$\Delta i_{DS} = f(v_{DS2}) - f(v_{DS1}), \tag{2.162}$$

we have

$$g_m \approx \frac{\Delta i_{DS}}{\Delta v_{GS}} = \frac{f(v_{DS2}) - f(v_{DS1})}{\Delta v_{GS}}. \tag{2.163}$$

g_m in this case increases with v_{DS} in a nonlinear fashion and will become nearly independent of v_{DS} once v_{DS} well passes the pinch-off voltage V_{sat}.

It was shown in [19] that the drain-modulated transconductance can be modeled approximately using

$$g_m(t) = \frac{1}{2} g_{m,max} \cos(\omega t), \tag{2.164}$$

where $g_{m,max}$ is the maximum transconductance. When an injection voltage

$$v_{inj} = V_{inj} \cos(\omega_{inj} t) \tag{2.165}$$

is applied to the gate of the transistor, the resultant channel current of the transistor will be obtained from

$$\begin{aligned} i_{DS} &= g_m(t) v_{GS} \\ &= \frac{1}{2} g_{m,max} V_{inj} \cos(\omega_{inj} t) \cos(\omega t). \end{aligned} \tag{2.166}$$

The mixing operation of the drain-pumped mixer is evident in (2.166).

If the injection signal applied to the gate of M3 is at 2ω, the corresponding channel current of M3 is given by

$$\begin{aligned} i_{DS3} &= g_{m3}(t)v_{GS3} \\ &= \frac{1}{2}g_{m,max}V_{inj}\cos(2\omega t)\cos(\omega t) \\ &= \frac{1}{4}g_{m,max}V_{inj}\left[\cos(\omega t)+\cos(3\omega t)\right]. \end{aligned} \tag{2.167}$$

A frequency component at ω is therefore generated and appears at the output of the oscillator, achieving a frequency division ratio of 2 [15, 45].

In [96], injection transistor M3 purposely operates in sub-threshold. In this case, a strong third-order harmonic component of the channel current of M3 will be observed when v_{DS3} varies. The drain-pumped operation will give rise to a large frequency component at 3ω. The transconductance of M3 can be written as

$$g_{m3}(t) = \frac{1}{2}g_{m,max}\cos(3\omega t). \tag{2.168}$$

If the injection signal applied to the gate is at 4ω, the corresponding channel current is given by

$$\begin{aligned} i_{DS} &= g_{m}(t)v_{GS} \\ &= \frac{1}{2}g_{m,max}V_{inj}\cos(4\omega t)\cos(3\omega t) \\ &= \frac{1}{4}g_{m,max}V_{inj}\left[\cos(\omega t)+\cos(7\omega t)\right]. \end{aligned} \tag{2.169}$$

A frequency component at ω is generated and appears at the output of the oscillator, achieving a frequency division ratio of 4.

2.6.4 Divide-by-3 Injection-Locked Frequency Dividers

It was shown in the preceding section that the complementary operation of the cross-coupled transistors M2 and M3 in Fig. 2.23 gives rise to the odd harmonics of the frequency of the oscillator. When an injection signal at 2ω is applied to the gate of the tail current source transistor, the current will be steered by M2 and M3 functioning as mixers that mix the current of the tail current source transistor with the gating signals of M2 and M3. The mixing operation generates various frequency components including the desired ω. The frequency component at ω passes through the LC resonators without attenuation while all other frequency components are suppressed by the LC resonators, thereby achieving a frequency division ratio of 2.

In order to realize a frequency division ratio of 3, the frequency of the injection signal should be 3ω while that of the output of the frequency divider needs to be ω. To achieve this, we need to generate a frequency component at 2ω and mix it with the injection signal at 3ω. Figure 2.27 shows the simplified schematic of the

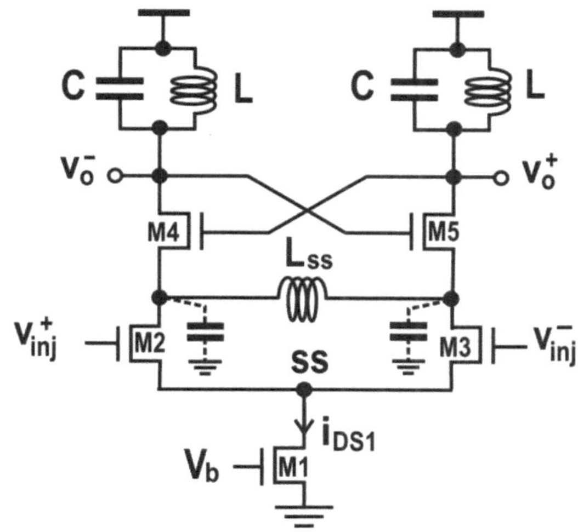

Fig. 2.27 Divide-by-3 ILFD [93]

divide-by-3 ILFD proposed by Wu and Zhang [93]. This frequency divider differs from the divide-by-2 ILFD studied earlier in Fig. 2.23 in the following aspects:

- The injection signal of the divide-by-3 ILFD is differentially injected into the source of cross-coupled transistors M4 and M5 that are in series with M2 and M3 while that of the divide-by-2 ILFD is injected into the common sources of transistors M2 and M3.
- Inductor L_{ss} is added between the drain of M2 and that of M3. It forms a parallel LC resonator with the capacitors encountered at the drains of M2 and M3. The value of L_{ss} is chosen such that the LC resonator resonates at 3ω.

To analyze the frequency divider, let

$$v_{inj}^+ = V_{inj,DC} + V_{inj}\cos(3\omega t), \tag{2.170a}$$

$$v_{inj}^- = V_{inj,DC} - V_{inj}\cos(3\omega t), \tag{2.170b}$$

where $V_{inj,DC}$ is the DC component of the injection signal and is needed to ensure that M2 and M3 operate in saturation. Differential injection ensures that the common-source node SS is an AC ground [65]. Since L_{ss} resonates out the capacitance encountered at the drains of M2 and M3, the load of M2 and M3 is resistive. As a result, the drain voltages of M2 and M3 are out of phase with the gate voltage of M2 and M3, respectively, and can be written as

$$v_{D2} = V_{DC} - V_m\cos(3\omega t), \tag{2.171a}$$

$$v_{D3} = V_{DC} + V_m\cos(3\omega t), \tag{2.171b}$$

where V_{DS} is the common-mode value of the drain voltages of M2 and M3.

Since inductor L_{ss} forms a parallel LC resonator with the capacitors at the drains of M2 and M3 and the LC resonator resonates at 3ω, it functions as a short circuit for frequency components at ω and an open-circuit for frequency components at 3ω. The effect of L_{ss} is thus two-fold:

1. It ensures that transistors M4 and M5 have the same source voltage at ω. Since v_o^+ and v_o^- are at ω, v_o^+ and v_o^- are therefore differential sinusoids of the same amplitude and frequency with a 50% duty cycle

$$v_o^+ = V_{o,DC} + V_{o,m}\cos(\omega t) \tag{2.172a}$$

$$v_o^- = V_{o,DC} - V_{o,m}\cos(\omega t), \tag{2.172b}$$

   where $V_{o,DC}$ is the common-mode voltage of v_o^+ and v_o^-.
2. When differential injection signals at 3ω are applied to the gates of M2 and M3, the frequency component of the resultant channel currents at 3ω will reach the sources of M4 and M5 without attenuation while all other frequency components of the channel current of M2 and M3 will be summed up, canceling out their AC components, as the drains of M2 and M3 collapse together at frequencies other than 3ω.

The injection of v_{inj}^+ and v_{inj}^- thus results in different voltages of frequency 3ω at the sources of M4 and M5. As a result, the operation of M4 and M5 is not differential. This differs from that of the divide-by-2 injection-locked frequency divider studied earlier in Fig. 2.23 where the operation of the cross-coupled transistors is differential at all frequencies. The mixing of frequency components at 3ω and those at ω is performed by M4 and M5 with frequency components at 3ω applied to the source and those at ω applied to the gate. The channel current of M4 and M5 is a nonlinear function of both v_{GS} and v_{DS} of the transistors. The nonlinearity of M4 and M5 gives rise to multiple frequency components including those at ω_o in their channel currents. Since the LC resonators resonate at ω_o only, only frequency components at ω_o will appear at v_o^+ and v_o^- while all other frequency components are shunted to the ground by the LC tanks. As a result, a frequency division of 3 is obtained.

The divide-by-3 ILFD shown in Fig. 2.27 is less attractive for low-voltage applications. This is because injection transistors M2 and M3 must operate in saturation. As a result, a large drain-source voltage drop of M2 and M3 is unavoidable. In [33, 93], the injection transistors in Fig. 2.27 are replaced with two transformers with the injection signals applied to the primary winding of the transformers. The secondary windings of the transformers are placed in the location where the injection transistors reside in Fig. 2.27, as shown in Fig. 2.28. Since the voltage drop across the secondary windings of the transformers in the DC steady state is zero, the minimum supply voltage is reduced. The removal of the

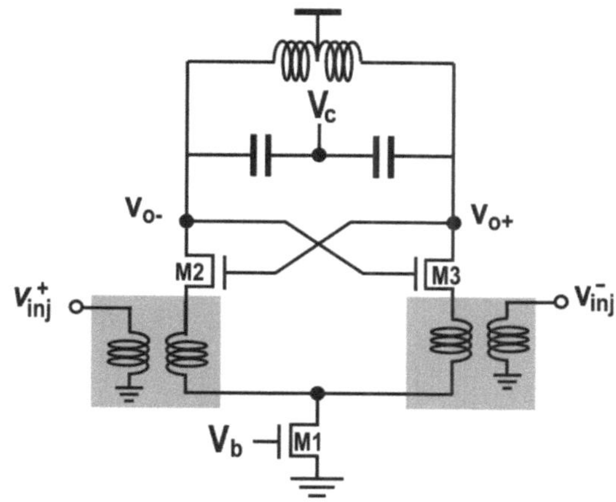

Fig. 2.28 Divide-by-3 ILFD with transformer injection

injection transistors also lowers the power consumption. The price paid for a low supply voltage and low power consumption is a large silicon area occupied by the transformers.

2.7 Subharmonic Injection

In this section, we study the subharmonic injection-locking of harmonic oscillators where the frequency of the injection signal is ω_o/N where $N = 2, 3, \ldots$ and ω_o is the free-running frequency of the oscillators under injection. Since the frequency of the injection signal is a fraction of that of the oscillator under injection, when the oscillator is locked to the injection signal, it will function as an integer frequency multiplier. The approach to be used to analyze the subharmonic injection-locking of harmonic oscillators is similar to that for analysis of the superharmonic injection-locking of harmonic oscillators, specifically, a harmonic oscillator in subharmonic injection is modeled as a nonlinear system with a nonlinear block depicting the nonlinear characteristics of the oscillator and a bandpass filter characterizing the frequency selection of the oscillator in the forward path and a unity-gain block in the feedback path. The nonlinear block performs frequency mixing so as to generate the desired frequency components, whereas the bandpass filter suppresses all other frequency components at the output of the nonlinear block and only allows those whose frequency falls into the passband of the bandpass filter to pass through and reach the output of the oscillator. Both second-order and third-order subharmonic injection-locking will be investigated in this section.

2.7.1 Second-Order Subharmonic Injection

Let us first consider that the frequency of the injection signal is approximately half that of the free-running frequency of the oscillator under injection

$$
\begin{aligned}
i_{inj}(t) &= I_{inj}\cos\left(\frac{\omega t}{2}+\phi_{inj}\right)\\
&= \frac{I_{inj}}{2}e^{j\frac{\omega t}{2}}e^{j\phi_{inj}}+\frac{I_{inj}}{2}e^{-j\frac{\omega t}{2}}e^{-j\phi_{inj}}.
\end{aligned} \tag{2.173}
$$

with ω in the vicinity of the free-running frequency of the oscillator. The block diagram of the oscillator under subharmonic injection is shown in Fig. 2.9 with the nonlinear block depicted by (2.66). When the oscillator is locked to the injection signal, its frequency will be ω. The output of the oscillator in the lock state can be written as

$$
\begin{aligned}
i_o &= I_o\cos(\omega t)\\
&= \frac{I_o}{2}e^{j\omega t}+\frac{I_o}{2}e^{-j\omega t}.
\end{aligned} \tag{2.174}
$$

The input and output of the nonlinear block are given by

$$
\begin{aligned}
i_{inj}(t) &= i_{inj}-i_o\\
&= \frac{I_{inj}}{2}e^{j\frac{\omega t}{2}}e^{j\phi_{inj}}+\frac{I_{inj}}{2}e^{-j\frac{\omega t}{2}}e^{-j\phi_{inj}}-\frac{I_o}{2}e^{j\omega t}-\frac{I_o}{2}e^{-j\omega t}.
\end{aligned} \tag{2.175}
$$

and

$$
\begin{aligned}
y &= a_1\left(\frac{I_{inj}}{2}e^{j\frac{\omega t}{2}}e^{j\phi_{inj}}+\frac{I_{inj}}{2}e^{-j\frac{\omega t}{2}}e^{-j\phi_{inj}}-\frac{I_o}{2}e^{j\omega t}-\frac{I_o}{2}e^{-j\omega t}\right)\\
&+a_2\left(\frac{I_{inj}}{2}e^{j\frac{\omega t}{2}}e^{j\phi_{inj}}+\frac{I_{inj}}{2}e^{-j\frac{\omega t}{2}}e^{-j\phi_{inj}}-\frac{I_o}{2}e^{j\omega t}-\frac{I_o}{2}e^{-j\omega t}\right)^2\\
&+a_3\left(\frac{I_{inj}}{2}e^{j\frac{\omega t}{2}}e^{j\phi_{inj}}+\frac{I_{inj}}{2}e^{-j\frac{\omega t}{2}}e^{-j\phi_{inj}}-\frac{I_o}{2}e^{j\omega t}-\frac{I_o}{2}e^{-j\omega t}\right)^3.
\end{aligned} \tag{2.176}
$$

It can be shown that among the frequency components at the output of the nonlinear block, only the following frequency components fall into the passband of the resonator and make their way to the output of the resonator:

- $-\frac{a_1}{2}I_oe^{j\omega t}$,
- $a_2\left(\frac{I_{inj}}{2}\right)^2e^{j2\phi_{inj}}e^{j\omega t}$, and
- $3a_3\left(-\frac{I_o}{2}\right)^2\left(-\frac{I_o}{2}\right)e^{j\omega t}$.

We have the following phasor domain relation at frequency ω

$$\left(-\frac{a_1}{2}I_o + \frac{a_2}{4}I_{inj}^2 e^{j2\phi_{inj}} - \frac{3a_3}{8}I_o^3\right)\frac{H_o}{1+j2Q\left(\frac{\omega-\omega_o}{\omega_o}\right)} = \frac{I_o}{2}. \tag{2.177}$$

Equating the real part of (2.177) yields

$$1 + a_1 H_o + \frac{3a_3}{4}H_o I_o^2 - \frac{a_2}{2}H_o\frac{I_{inj}^2}{I_o}\cos(2\phi_{inj}) = 0. \tag{2.178}$$

To derive lock range

$$\Delta\omega_{L,2nd,sub} = \omega - \omega_o, \tag{2.179}$$

we equate the imaginary part of (2.177) gives

$$\Delta\omega_{L,2nd,sub} = \frac{\omega_o}{2Q}\left(\frac{a_2}{2}H_o\right)\frac{I_{inj}^2}{I_o}\sin(2\phi_{inj}). \tag{2.180}$$

It is seen from (2.180) that the larger the injection signal, the larger the lock range. Also, the stronger the second-order nonlinearity signified by a_2, the larger the lock range. Moreover, the larger the gain of the resonator H_o at the center of the passband of the resonator, the larger the lock range for reasons similar to superharmonic injections investigated earlier. It is also seen that the smaller the output of the oscillator, the larger the lock range. The reason for this distinct characteristic is that when the output of the oscillator is large, the two components $-\frac{a_1}{2}I_o e^{j\omega t}$ and $-\frac{3a_3}{4}I_o^3 e^{j\omega t}$ at the output of the nonlinear block will be large. They will reduce the net signal at ω that is fed to the resonator, resulting in a smaller frequency displacement.

The lock range is also dependent of injection phase ϕ_{inj}. The lock range will be maximized when

$$\phi_{inj}^* = \frac{\pi}{4}. \tag{2.181}$$

The maximum lock range is given by

$$\Delta\omega_{L,2nd,sub,max} = \frac{\omega_o}{2Q}\left(\frac{a_2}{2}H_o\right)\frac{I_{inj}^2}{I_o}. \tag{2.182}$$

Figure 2.29 plots i_{inj} and i_o with $\phi_{inj} = \pi/4$. ϕ_{inj} is measured with reference to $\omega/2$. It is seen that injection signal is large when i_o crosses the threshold. For oscillators in superharmonic injection, i_{inj} peaks at every threshold-crossing of i_o. In subharmonic injection, i_{inj} cannot peak at every threshold-crossing of i_o. $\phi_{inj} = \pi/4$ is the case where i_{inj} is large at every threshold-crossing. Should the injection phase deviates from $\pi/4$, such as $\phi_{inj} = \pi/6$ and $\phi_{inj} = \pi/3$, i_{inj} is only large at some threshold-crossings while small at others, as shown in the figure.

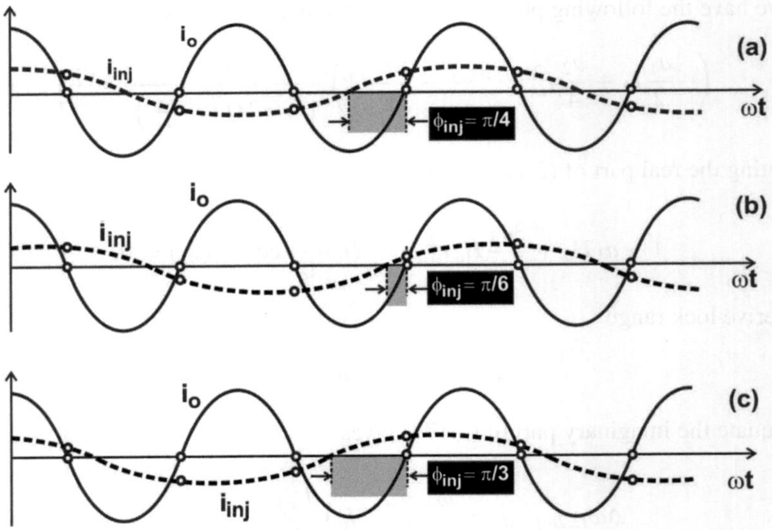

Fig. 2.29 Second-order subharmonic injection with i_{inj} leading i_o by **(a)** $\pi/4$, **(b)** $\pi/6$, and **(c)** $\pi/3$

2.7.2 Third-Order Subharmonic Injection

Let us now consider that the frequency of the injection signal is approximately one-third that of the free-running frequency of the oscillator under injection

$$\begin{aligned} i_{inj}(t) &= I_{inj} \cos\left(\frac{\omega t}{3} + \phi_{inj}\right) \\ &= \frac{I_{inj}}{2} e^{j\frac{\omega t}{3}} e^{j\phi_{inj}} + \frac{I_{inj}}{2} e^{-j\frac{\omega t}{3}} e^{-j\phi_{inj}}, \end{aligned} \tag{2.183}$$

and 3ω is in the vicinity of the free-running frequency of the oscillator under injection. When the oscillator is locked to the injection signal, its frequency will be ω. The output of the oscillator in the lock state is given by

$$\begin{aligned} i_o &= I_o \cos(\omega t) \\ &= \frac{I_o}{2} e^{j\omega t} + \frac{I_o}{2} e^{-j\omega t}. \end{aligned} \tag{2.184}$$

The input of the nonlinear block is given by

$$\begin{aligned} x &= i_{inj} - i_o \\ &= \frac{I_{inj}}{2} e^{j\frac{\omega t}{3}} e^{j\phi_{inj}} + \frac{I_{inj}}{2} e^{-j\frac{\omega t}{3}} e^{-j\phi_{inj}} - \frac{I_o}{2} e^{j\omega t} - \frac{I_o}{2} e^{-j\omega t}, \end{aligned} \tag{2.185}$$

from which we obtain the output of the nonlinear block

$$\begin{aligned} y &= a_1 x + a_2 x^2 + a_3 x^3 \\ &= a_1 \left(\frac{I_{inj}}{2} e^{j\frac{\omega t}{3}} e^{j\phi_{inj}} + \frac{I_{inj}}{2} e^{-j\frac{\omega t}{3}} e^{-j\phi_{inj}} - \frac{I_o}{2} e^{j\omega t} - \frac{I_o}{2} e^{-j\omega t} \right) \\ &\quad + a_2 \left(\frac{I_{inj}}{2} e^{j\frac{\omega t}{3}} e^{j\phi_{inj}} + \frac{I_{inj}}{2} e^{-j\frac{\omega t}{3}} e^{-j\phi_{inj}} - \frac{I_o}{2} e^{j\omega t} - \frac{I_o}{2} e^{-j\omega t} \right)^2 \\ &\quad + a_3 \left(\frac{I_{inj}}{2} e^{j\frac{\omega t}{3}} e^{j\phi_{inj}} + \frac{I_{inj}}{2} e^{-j\frac{\omega t}{3}} e^{-j\phi_{inj}} - \frac{I_o}{2} e^{j\omega t} - \frac{I_o}{2} e^{-j\omega t} \right)^3 . \end{aligned} \tag{2.186}$$

It can be shown that among the frequency components at the output of the nonlinear block, only the following frequency components fall into the passband of the resonator and make their way to the output of the resonator:

- $-\frac{a_1 I_o}{2} e^{j\omega t}$,
- $3a_3 \left(\frac{I_{inj}}{2}\right)^3 e^{j\omega t + j\phi_{inj}}$, and
- $3a_3 \left(-\frac{I_o}{2}\right)^2 \left(-\frac{I_o}{2}\right) e^{j\omega t}$.

We therefore arrive at the following phasor domain relation at frequency ω

$$\left(-\frac{a_1}{2} I_o + \frac{a_3}{8} I_{inj}^3 e^{j3\phi_{inj}} - \frac{3a_3}{8} I_o^3 \right) \frac{H_o}{1 + j2Q\left(\frac{\Delta\omega_L}{\omega_o}\right)} = \frac{I_o}{2}. \tag{2.187}$$

Equating the real part of (2.187) gives

$$1 + a_1 H_o + \frac{3a_3}{4} H_o I_o^2 - \frac{a_3}{4} H_o \frac{I_{inj}^3}{I_o} \cos(3\phi_{inj}) = 0. \tag{2.188}$$

Equating the imaginary part of (2.187) yields

$$\Delta\omega_{L,3rd,sub} = \frac{\omega_o}{2Q} \left(\frac{a_3}{4} H_o \right) \frac{I_{inj}^3}{I_o} \sin(3\phi_{inj}). \tag{2.189}$$

It is seen from (2.189) that the larger the injection signal, the larger the lock range. Also, the stronger the third-order nonlinearity signified by a_3, the larger the lock range. Further, the larger the gain of the resonator H_o at the center of the passband of the resonator, the larger the lock range. Moreover, the smaller the output of the oscillator, the larger the lock range. The reason for this distinct characteristic is that when the output of the oscillator is large, the two components $-\frac{a_1}{2} I_o e^{j\omega t}$ and $-\frac{3a_3}{8} I_o^3 e^{j\omega t}$ at the output of the nonlinear block will be large. They will reduce the net signal at ω that is fed to the resonator, resulting in a smaller frequency displacement.

The lock range is also dependent of injection phase ϕ_{inj}. The lock range will be maximized when

$$\phi_{inj} = \frac{\pi}{6}. \tag{2.190}$$

The maximum lock range is given by

$$\Delta\omega_{L,3rd,sub,max} = \frac{\omega_o}{2Q}\left(\frac{a_3}{4}H_o\right)\frac{I_{inj}^3}{I_o}. \tag{2.191}$$

Figure 2.30 shows i_{inj} and i_o with $\phi_{inj} = \pi/6$. Note that ϕ_{inj} is measured with reference to $\omega/3$. It is seen that i_{inj} is large at every threshold-crossing of i_o, ensuring the maximum impact of the injection signal on the frequency of the oscillator. Should the injection phase deviate from the optimal phase such as $\phi_{inj} = \pi/3$ and $\phi_{inj} = \pi/4$, i_{inj} will be large only at some threshold-crossings while small at others, resulting in a reduced lock range.

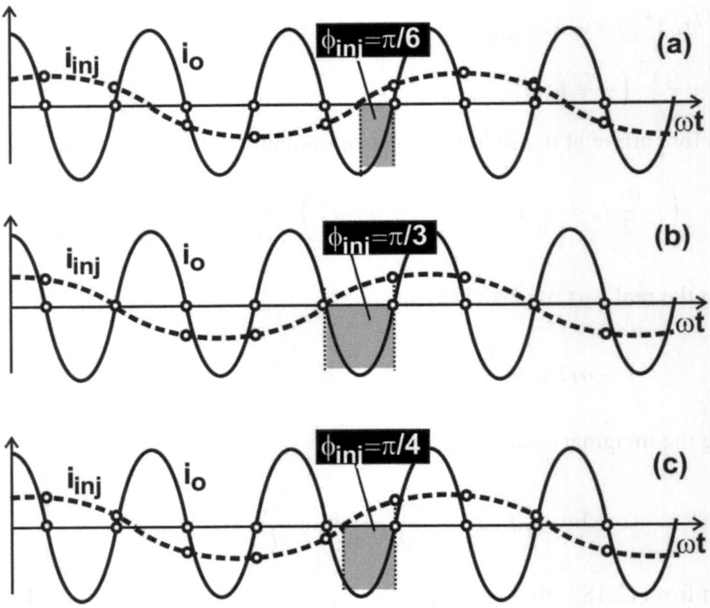

Fig. 2.30 Third-order subharmonic injection with i_{inj} leading i_o by **(a)** $\pi/6$, **(b)** $\pi/3$, and **(c)** $\pi/4$

2.8 Lock Range Relation

In the preceding sections, we derived the lock range of harmonic oscillators in first-harmonic, superharmonic, and subharmonic injections. Their different lock ranges are rooted to the distinct characteristics of injection signals. Since the oscillator under injection is the same, intrinsic relations between the lock range of harmonic oscillators in first-harmonic, superharmonic, and subharmonic injections must exist. In this section, we explore these intrinsic relations.

To facilitate analysis, the maximum lock range of harmonic oscillators in first-harmonic, superharmonic, and subharmonic injections derived earlier are tabulated in Table 2.2 for convenience.

2.8.1 Lock Range Relation: Superharmonic Injection

It can be shown from Table 2.2 that the ratio of the maximum lock range of a harmonic oscillator in second-order superharmonic injection to that in first-harmonic injection with mild third-order nonlinearity, i.e., $a_3 \ll a_1$ is given by

$$\left|\frac{\Delta\omega_{L,2nd,super,max}}{\Delta\omega_{L,1st,max}}\right| \approx \left|\frac{a_2}{a_1}\right| I_o. \tag{2.192}$$

It is seen from (2.192) that the lock range ratio is not only proportional to a_2, which signifies the second-order nonlinearity of the oscillator, but also proportional to the output of the oscillator. The reason for the former is straightforward while that for the latter is that the output of the oscillator is fed to the nonlinear block of the oscillator directly. The larger the output of the oscillator at ω, the larger the second-order intermodulation component of the output of the nonlinear block generated by the input at 2ω and the output at ω subsequently the larger the input of the resonator

Table 2.2 The maximum lock range of harmonic oscillators in first-harmonic, superharmonic, and subharmonic injections with mild third-order nonlinearity ($a_3 \ll a_1$)

Injection	Max. lock range
First-harmonic	$\left\|\Delta\omega_{L,dir,max}\right\| \approx \frac{\omega_o}{2Q}\frac{I_{inj}}{I_o}(a_1 H_o)$
Second-order superharmonic	$\left\|\Delta\omega_{L,2nd,super,max}\right\| = \frac{\omega_o}{2Q}(a_2 H_o) I_{inj}$
Third-order superharmonic	$\left\|\Delta\omega_{L,3rd,super,max}\right\| = \frac{\omega_o}{2Q}\left(\frac{3a_3}{4} H_o\right) I_{inj} I_o$
Second-order subharmonic	$\left\|\Delta\omega_{L,2nd,sub,max}\right\| = \frac{\omega_o}{2Q}\left(\frac{a_2}{2} H_o\right)\frac{I_{inj}^2}{I_o}$
Third-order subharmonic	$\left\|\Delta\omega_{L,3rd,sub,max}\right\| = \frac{\omega_o}{2Q}\left(\frac{a_3}{4} H_o\right)\frac{I_{inj}^3}{I_o}$

at ω. Since the loop gain at ω imposed by Barkhausen magnitude criterion needs to be unity, a large frequency displacement from the center frequency of the passband of the resonator is required so as to have a sufficient drop in the magnitude of the transfer function of the resonator in order to satisfy Barkhausen magnitude criterion.

Similarly, one can find the ratio of the maximum lock range of the oscillator in third-order superharmonic injection to that in first-harmonic injection

$$\left|\frac{\Delta\omega_{L,3rd,super,max}}{\Delta\omega_{L,1st,max}}\right| = \frac{3}{4}\left|\frac{a_3}{a_1}\right| I_o^2. \tag{2.193}$$

Equation (2.193) shows that the lock range ratio is not only proportional to a_3, which signifies the third-order nonlinearity of the oscillator, but also proportional to the square of the output of the oscillator. The reason for the latter is that the larger the output of the oscillator at ω, the larger the frequency component at 2ω generated by squaring the output of the oscillator prior to its mixing with the injection signal at 3ω and the larger the input of the resonator at ω. As the loop gain at ω must be unity, a large frequency displacement from the center frequency of the passband of the resonator is therefore required in order to have an adequate drop in the magnitude of the transfer function of the resonator so as to satisfy Barkhausen magnitude criterion.

2.8.2 Lock Range Relation: Subharmonic Injection

The ratio of the maximum lock range of the oscillator in second-order subharmonic to that in first-harmonic injection is given by

$$\left|\frac{\Delta\omega_{L,2nd,sub,max}}{\Delta\omega_{L,1st,max}}\right| = \frac{1}{2}\frac{a_2}{a_1} I_i. \tag{2.194}$$

It is seen from (2.194) that the lock range ratio is not only proportional to a_2, but also proportional to the strength of the injection signal. The reason for the latter is that the larger the injection signal at $\omega/2$, the larger the frequency component at ω at the output of the nonlinear block subsequently the larger the input of the resonator. Since the loop gain at ω imposed by Barkhausen magnitude criterion needs to be unity, a large frequency displacement from the center frequency of the passband of the resonator is required so as to have the needed drop in the magnitude of the transfer function of the resonator.

Similarly, one can show

$$\left|\frac{\Delta\omega_{L,3rd,sub,max}}{\Delta\omega_{L,1st,max}}\right| = \frac{1}{4}\frac{a_3}{a_1} I_i^2. \tag{2.195}$$

Equation (2.195) shows that the ratio is not only proportional to a_3, but also proportional to the square of the input of the oscillator. The reason for the latter is that the larger the injection signal at $\omega/3$, the larger its third-order harmonic component at the output of the nonlinear block at ω subsequently the larger the input of the resonator. Since the loop gain at ω imposed by Barkhausen magnitude criterion needs to be unity, a large frequency displacement from the center frequency of the passband of the resonator is required so as to have the needed drop in the magnitude of the transfer function of the resonator.

The preceding results confirm the existence of intrinsic relations among the lock range of harmonic oscillators in first-harmonic, superharmonic, and subharmonic injection.

2.9 Phase Noise of Injection-Locked Harmonic Oscillators

In Sect. 2.3, we investigated the phase noise of free-running harmonic oscillators. In this section, we explore the phase noise of harmonic oscillators that are locked to an injection signal whose phase noise is negligible as compared with that of the free-running oscillators.

Consider the harmonic oscillator under the injection of i_{inj} shown in Fig. 2.31. Let S_{in} be the PSD of the noise of the channel current of the transistor. It was shown earlier that the injection-locked harmonic oscillator can be represented by the linear feedback system shown in the figure if the transistor is considered as a linear transconductor. When the injection signal is absent, the oscillator will oscillate at its natural frequency ω_o and Barkhausen criteria holds at ω_o, i.e., $1 + H_o(\omega_o) = 0$. When an injection signal of frequency ω_{inj} is present and ω_{inj} falls into the lock range of the oscillator, the injection signal will cause the frequency of the oscillator to change from its natural frequency ω_o to that of the injection signal $\omega_{inj} = \omega_o + \Delta\omega$ with $\Delta\omega \ll \omega_o$. The closed-loop transfer function is obtained from

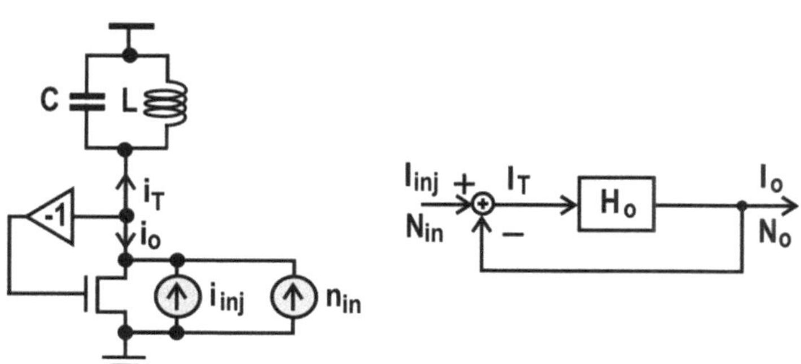

Fig. 2.31 Analysis of the phase noise of injection-locked harmonic oscillators

$$H_c(\omega_o+\Delta\omega)=\frac{H_o(\omega_o+\Delta\omega)}{1+H_o(\omega_o+\Delta\omega)} \approx \frac{-1}{\left(\frac{dH_o}{d\omega}\right)_{\omega_o}(\Delta\omega)}. \quad (2.196)$$

Making use of the quality factor of oscillators introduced in (2.37), we arrive from (2.196)

$$|H_c(\omega_o+\Delta\omega)|^2 \approx \frac{1}{4Q^2}\left(\frac{\omega_o}{\Delta\omega}\right)^2. \quad (2.197)$$

The PSD of the output noise of the free-running oscillator at frequency $\omega_{in} = \omega_o + \Delta\omega$, denoted by $S_{o,free}$, is given by

$$S_{o,free}(\omega_{inj}) = |H_c(\omega_{inj})|^2 S_{in}(\omega_{inj}) \approx \frac{1}{4Q^2}\left(\frac{\omega_o}{\Delta\omega}\right)^2 S_{in}(\omega_{inj}). \quad (2.198)$$

When unitone injection signal i_{inj} at ω_{inj} is presented, the oscillator will lock to the injection signal and oscillate at ω_{inj}. Let the injection signal and the output of the injection-locked oscillator be given by

$$i_{inj} = I_{inj}e^{j\omega_{inj}t}, \quad (2.199a)$$

$$i_o = I_o e^{j(\omega_{inj}t+\phi_o)}, \quad (2.199b)$$

respectively where ϕ_o is the phase of the output of the oscillator. Note that the phase of the injection signal is chosen to be the phase reference. Since the closed-loop transfer function of the injection-locked oscillator is given by [50]

$$|H_c(\omega_{inj})| = \left|\frac{I_o}{I_{inj}}\right|, \quad (2.200)$$

the PSD of the output noise of the injection-locked oscillator at frequency $\omega_{in} = \omega_o + \Delta\omega$, denoted by $S_{o,inj}$, is obtained from

$$S_{o,inj}(\omega_{inj}) = \left|\frac{I_o}{I_{inj}}\right|^2 S_{in}(\omega_{inj}). \quad (2.201)$$

The ratio of (2.201) to (2.198) gives

$$\frac{S_{o,inj}(\omega_{inj})}{S_{o,free}(\omega_{inj})} \approx 4Q^2\left|\frac{I_o}{I_{inj}}\right|^2\left(\frac{\Delta\omega}{\omega_o}\right)^2. \quad (2.202)$$

The maximum lock range of the oscillator was derived in (2.87) earlier and is repeated here for convenience

$$\left|\Delta\omega_{L,max}\right| = \frac{\omega_o}{2Q}\left|\frac{I_{inj}}{I_o}\right|. \tag{2.203}$$

Making use of (2.203), we arrive from (2.202)

$$S_{o,inj}(\omega_{inj}) \approx \left|\frac{\Delta\omega}{\Delta\omega_{L,max}}\right|^2 S_{o,free}(\omega_{inj}). \tag{2.204}$$

Let us examine (2.204) in detail:

(a) When the frequency of the injection signal ω_{inj} is close to the free-running frequency of the oscillator ω_o, i.e., $\Delta\omega = \omega_{inj} - \omega_o \ll \Delta\omega_{L,max}$, $S_{o,inj}(\omega_{inj}) \ll S_{o,free}(\omega_{inj})$, the phase noise of the injection-locked oscillator will be much smaller as compared with that of the free-running oscillator.
(b) When injection frequency ω_{inj} deviates from the free-running frequency of the oscillator, $S_{o,inj}(\omega_{inj})$ will rise and approach to $S_{o,free}(\omega_{inj})$.
(c) When injection frequency $\omega_{inj} - \omega_o = \Delta\omega_{L,max}$, $S_{o,inj}(\omega_{inj}) = S_{o,free}(\omega_{inj})$ will hold. The phase noise of the injection-locked harmonic oscillator is the same as that of the free-running oscillator when the frequency of the injection signal is at the edges of the lock range.

The preceding observations reveal that in order to minimize the phase noise of an injection-locked harmonic oscillator, two factors must be considered: (a) The maximum lock range $\Delta\omega_{L,max}$ of the oscillator should be made as large as possible. A larger $\Delta\omega_{L,max}$ not only ameliorates the success of injection-locking, but also lowers the phase noise of the oscillator once locked to the injection signal. (b) The frequency of the clean injection signal should be as close as possible to the natural frequency of the oscillator.

2.10 Chapter Summary

The fundamentals of the oscillation of harmonic oscillators were studied. The quality factor of harmonic oscillators was introduced and the spectrum of the phase noise of harmonic oscillators was investigated. An emphasis was given to the up-conversion of the flicker noise of the tail current source transistor of oscillators. The modeling of harmonic oscillators was investigated. We showed that a harmonic oscillator can be depicted using a nonlinear system with a nonlinear block characterizing the nonlinear characteristics of the oscillator and a high-Q bandpass filter signifying the frequency selectivity of the oscillator.

For first-harmonic injection, we showed that the lock range of a harmonic oscillator is a function of both the strength of the injection signal and the degree of the nonlinearity of the oscillator. We further showed that the impact of the injection signal can be represented by the variation of the impedance of the resonators such that the injection-locked oscillator can be treated as an autonomous system with the injection signal represented by the variation of the impedance of the resonator. Such a treatment enables the quantification of the impact of the injection-induced variation of the impedance of the resonator so as to satisfy Barkhausen criteria. Harmonic oscillators with an active inductor resonator exhibit a larger lock range as compared with those with a spiral inductor resonator due to the injection-induced variation of the inductance of the active inductors and the low-quality factor of the active inductor.

For second-order superharmonic injection, we showed that lock range is proportional to both the degree of the second-order nonlinearity of oscillators and the strength of the injection signal. For harmonic oscillators in third-order superharmonic injection, we showed that the lock range is proportional to the degree of the third-order nonlinearity of oscillators, the strength of injection signal, and the output of the oscillator. The operation of divide-by-2 and divide-by-3 injection-locked frequency dividers was detailed. We showed divide-by-2 ILFDs with tail injection suffer from a low injection efficiency due to the loss of the injection signal at the common-source node. Injection efficiency can be improved using drain-pumped mixing. As these frequency dividers rely on the operation of the injection transistor in saturation region just outside the pinch-off, careful biasing is needed. Divide-by-3 ILFDs with a pair of differentially configured injection transistors are not particularly attractive for low-voltage applications due to the large drain-source voltage of these transistors. This drawback can be eliminated using spiral transformer injection, however, at the cost of a significantly increased silicon area.

For second-order subharmonic injection, we showed that lock range is proportional to both the degree of the second-order nonlinearity of oscillators and the square of the injection signal. For harmonic oscillators in third-order subharmonic injection, we showed that the lock range is proportional to the degree of the third-order nonlinearity of oscillators and the cube of the injection signal.

The intrinsic relations between the lock range of harmonic oscillators in first-harmonic, superharmonic, and subharmonic injections were derived. The basis of these intrinsic relations is that the oscillator under injection is the same while its injection signals differ.

Finally, the phase noise of injection-locked harmonic oscillators was investigated. We showed that when the frequency of the clean injection signal of a harmonic oscillator is close to the free-running frequency of the oscillator, the phase noise of the injection-locked oscillator will be much lower as compared with that of the free-running oscillator. When injection frequency moves away from the free-running frequency of the oscillator, the phase noise of the injection-locked oscillators will deteriorate. When the frequency of the injection signal is at the boundaries of the lock range, the phase noise of the injection-locked oscillator will be on par with that of the free-running oscillator. The lock range of oscillators should therefore be maximized so as to not only ameliorate the success of injection-locking but also lower the phase noise of the oscillators once locked. The frequency of the clean injection signal should also be as close as possible to the natural frequency of the oscillator under injection so as to minimize the phase noise of the oscillator once locked.

Chapter 3
Injection-Locking Techniques for Harmonic Oscillators

3.1 Introduction

A large lock range is critical for applications such as ILFDs and injection-lock based frequency calibration of PWMs. ILFDs are often used as the pre-scalar of a frequency synthesizer to take the advantages of the high operation frequency and low power consumption of ILFDs [19, 93, 94, 96]. Harmonic oscillators are typically used in ILFD pre-scalars for wireless applications, attributed to their better phase noise performance as compared with their nonharmonic counterparts. Because not only the frequency of the oscillator of the frequency divider is subject to the impact of PVT uncertainty, the frequency of the input of the frequency divider, which is the output of the oscillator of the frequency synthesizer, is also subject to the effect of PVT uncertainty. As a result, the frequency of the input of the ILFD pre-scalar might fall outside the lock range of the oscillator of the ILFD. Increasing the lock range of ILFDs is clearly critical. For PWMs, the system clock of a PWM can be generated using a local oscillator residing in the PWM so as to avoid the drawbacks of generating the system clock of the PMW directly from its carrier [37]. The frequency of the local oscillator is sensitive to the effect of PVT uncertainty and must be calibrated against a timing reference sent from the base station of the PWM prior to its intended operations. Although a digital FLL can serve the purpose of frequency calibration, digital FLLs suffer from a long lock time, a low frequency accuracy, and a large amount of power consumption. Injection-locking, on the other hand, offers fast locking, a high frequency accuracy, and low power consumption, and is therefore well suitable for the frequency calibration of PWMs. A large lock range is pivotal to the success of the injection-lock based frequency calibration of PWMs.

In Chap. 2, we showed that the stronger the injection signal, the larger the lock range. Strengthening the external injection signal of a harmonic oscillator, however, might not always result in a larger lock range. This is because the effective injection signal, which is the signal fed to the core of a harmonic oscillator, such as the

F. Yuan, *Injection-Locking in Mixed-Mode Signal Processing*,
https://doi.org/10.1007/978-3-030-17364-7_3

nonlinear core of a frequency divider where the desired frequency component is generated, might be weak due to the loss of the signal along the path through which the external injection signal propagates from the injection node to the core of the oscillator even though the external injection signal is large. The injection-locking of a harmonic oscillator is the result of the interaction between the injection signal and the oscillator, the characteristics of the injection signal such as the amplitude and frequency of the injection signal, the characteristics of the oscillator such as the quality factor and gain of the resonator of the oscillator, the degree of the nonlinearity of the core of the oscillator, and the transfer function from the injection node to the core of the oscillator where the desired frequency component is generated.

This chapter explores the factors that affect the lock range of harmonic oscillators and the techniques that increase the lock range of harmonic oscillators. Our focus is on ILFDs, which are the most representative applications of injection-locked harmonic oscillators. The remainder of the chapter is organized as follows: Sect. 3.2 investigates the dependence of the lock range of harmonic oscillators on injection signaling. Specifically, we investigate the lock range of harmonic oscillators with two unitone injections that are injected into two symmetrical locations of the oscillators and the relation between the lock range and the phase of the injection signals. Section 3.3 studies inductive series-peaking that increases the lock range of divide-by-2 and divide-by-3 ILFDs. Section 3.4 explores inductive shunt-peaking that increases the lock range of divide-by-2 ILFDs with tail injection. Section 3.5 deals with current-reuse direct injection in divide-by-2 ILFDs. In Sect. 3.6, our focus is on lock range improvement via lowering the quality factor of the resonator of ILFDs. Section 3.7 examines resistor feedback that boosts the lock range of harmonic oscillators without sacrificing the quality factor of the oscillator. Section 3.8 explores techniques that exact the second harmonic of the output of harmonic oscillators so as to realize divide-by-3 ILFDs. Section 3.9 studies transformer feedback that increases the lock range of ILFDs. Dual injections that increase the lock range of ILFDs are explored in Sect. 3.10. The chapter is summarized in Sect. 3.11.

3.2 Injection Signaling

The injection-locking of a harmonic oscillator by an external signal is an interaction process between the injection signal and the oscillator. Whether the oscillator will lock to the injection signal or not depends upon the characteristics of the injection signal such as the amplitude, frequency, and phase of the injection signal, and those of the oscillator, such as the degree of the nonlinearity of the oscillator, the transfer function of the path through which the external injection signal propagates from the terminal of the injection to the core of the oscillator where the desired frequency component is generated, and the gain of the resonator at its passband center frequency. In this section, we investigate the impact of the characteristics of injection signals on the lock range of harmonic oscillators. Such a study is well warranted as in many applications, both the oscillators under injection and the location to which the injection signal is applied, cannot be modified. Designers only have the freedom to orchestra injection signals in order to maximize lock range.

In Chap. 2, we investigated the lock range of harmonic oscillators with a unitone injection signal, i.e., the injection signal is a sinusoid. We showed that the lock range of a harmonic oscillator with a unitone injection is directly proportional to the relative strength of the injection signal and inversely proportional to the quality factor of the oscillator. In this section, we investigate harmonic oscillators with multiple unitone injections. We want to find out whether multiple unitone injections increase lock range or not. If they do, we further want to know whether the phase of the injection signals has any impact on the lock range or not. If it does, what is the relation between the lock range and the phase of the injection signals and what is the optimal phase of the injection signals that yields the maximum lock range? To simplify analysis, only the lock range of harmonic oscillators with two unitone injection signals of the same frequency and strength are considered. The underlined principle is valid for harmonic oscillators with more than two unitone injection signals.

Consider the LC oscillator with two unitone injection currents i_{inj1} and i_{inj2} shown in Fig. 3.1. The free-running frequency of the oscillator is given by $\omega_o = \frac{1}{\sqrt{LC}}$. Currents i_{inj1} and i_{inj2} are two sinusoids of the same frequency ω_{inj} and amplitude I_{inj} but different phases ϕ_{inj1} and ϕ_{inj2}, respectively. ϕ_{inj1} and ϕ_{inj2} are defined with respect to the phase of the output of the oscillator, which is chosen to be the reference phase

$$i_{inj1} = I_{inj} e^{j(\omega_{inj}t+\phi_{inj1})}, \tag{3.1a}$$

$$i_{inj2} = I_{inj} e^{j(\omega_{inj}t+\phi_{inj2})}. \tag{3.1b}$$

Equation (3.1) can be written as ,

$$i_{inj1} = I_{inj1} e^{j\omega_{inj}t}, \tag{3.2a}$$

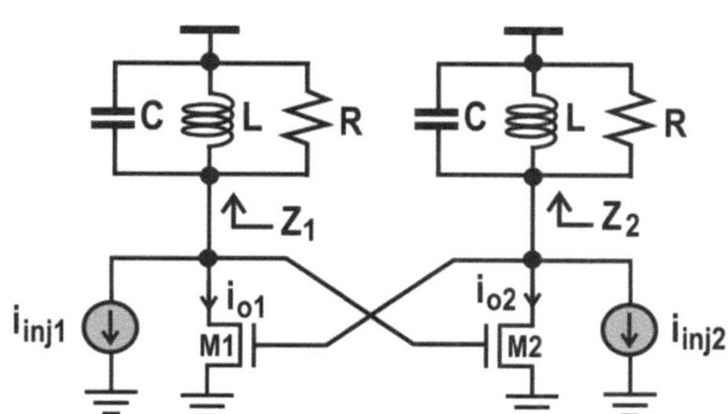

Fig. 3.1 Injection-locked harmonic oscillator with two unitone injection signals of the same frequency and amplitude

$$i_{inj2} = I_{inj2}e^{j\omega_{inj}t}, \tag{3.2b}$$

where

$$I_{inj1} = I_{inj}e^{j\phi_1}, \tag{3.3a}$$

$$I_{inj2} = I_{inj}e^{j\phi_2}, \tag{3.3b}$$

are the phasor of i_{inj1} and i_{inj2}, respectively.

Assume that ω_{inj} is within the lock range of the oscillator for given injection strength I_{inj}. The oscillator will oscillate at ω_{inj} once it is locked to the injection signals. Since in the lock state, only frequency ω_{inj} exists at both the input and output of the oscillator despite the existence of frequencies other than ω_{inj} inside the oscillator, arising from the nonlinearity of the oscillator, the oscillator can be represented by the linear system shown in Fig. 3.2 where g_{m1} and g_{m2} are the transconductances of M1 and M2, respectively. Note that the two transconductors are assumed to be ideal, i.e., they have an infinite input resistance, a zero output resistance, and no capacitance. Such a treatment is perfectly valid as the output resistance of M1 and M2 are absorbed into R, along with the series resistance of the spiral. Similarly, the capacitances of M1 and M2 are absorbed into C along with the parasitic capacitances of the spirals. Note that the direction of the output current of the transconductors ensures that the total phase shift around the loop is $-360°$ at the resonant frequency of the RLC tanks. This is because the increase of V_1 will lead to the decrease of V_2 as

$$V_2 = -Z_2 I_{o2} = -Z_2 g_{m2} V_1 \tag{3.4}$$

if no injection. Similarly, the decrease of V_2 leads to the increase of V_1 as

$$V_1 = -Z_1 I_{o1} = -Z_1 g_{m1} V_2. \tag{3.5}$$

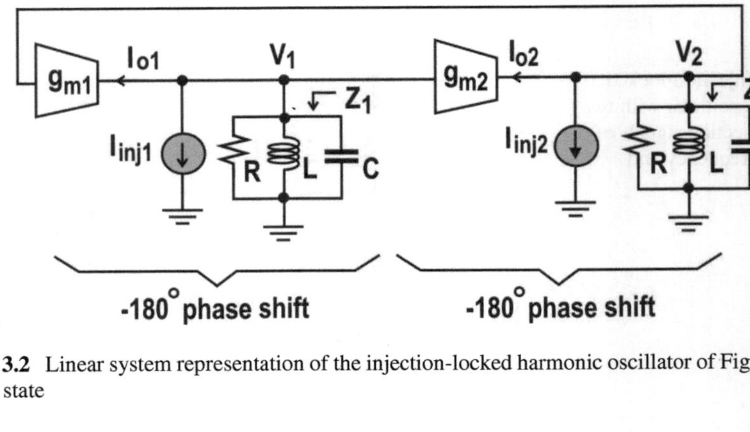

Fig. 3.2 Linear system representation of the injection-locked harmonic oscillator of Fig. 3.1 in the lock state

Since at the self-resonant frequency of the RLC tanks, $Z_1 = Z_2 = R$, each stage indeed has $-180°$ phase shift. The total phase shift of $-360°$ around the loop is therefore obtained. The $-180°$ of each stage also ensures that V_1 and V_2 are out of phase or their phase difference is $180°$.

It should be emphasized that Fig. 3.2 is valid for the oscillator in the lock state only. It does not account for frequency components other than ω_{inj} that do exist inside the oscillator but vanish prior to reaching the output of the oscillator. As a result, Fig. 3.2 cannot provide information such as the generation, propagation, and diminishing of frequency components other than ω_{inj} inside the oscillator. Such information is often critical to our understanding of both the locking process of the oscillator and the mechanisms that determine the lock range of the oscillator, as to be seen in the following sections. Since Fig. 3.2 quantifies the behavior of the oscillator in the lock state, it is capable of yielding the lock range of the oscillator, as to be seen shortly.

When the injection signals are absent, i.e., $i_{inj1}, i_{inj2} = 0$, we will have

$$V_1 = -Z_1(\omega_o)I_{o1}, \tag{3.6a}$$

$$V_2 = -Z_2(\omega_o)I_{o2}, \tag{3.6b}$$

where

$$Z_1(\omega_o) = R_1||j\omega_o L_1||\frac{1}{j\omega_o C_1} \tag{3.7a}$$

$$Z_2(\omega_o) = R_2||j\omega_o L_2||\frac{1}{j\omega_o C_2} \tag{3.7b}$$

are the impedance of the resonators at their self-resonant frequency. Note that the variables in (3.6) are evaluated at ω_o.

When the injection signals are present and their frequency falls into the lock range of the oscillator, the oscillator will be locked to the injection signal and oscillate at ω_{inj}. We have

$$V_1 = -Z_1(I_{o1} + I_{inj1}), \tag{3.8a}$$

$$V_2 = -Z_2(I_{o2} + I_{inj2}). \tag{3.8b}$$

Note that the variables in (3.8) are evaluated at ω_{inj}. Because

$$I_{o1} = g_{m1}\left[-\left(I_{o2} + I_{inj2}\right) Z_2\right]. \tag{3.9a}$$

$$I_{o2} = g_{m2}\left[-\left(I_{o1} + I_{inj1}\right) Z_1\right], \tag{3.9b}$$

Solve for I_{o1} and I_{o2}

$$I_{o1} = \frac{g_{m1}g_{m2}Z_1Z_2}{1 - g_{m1}g_{m2}Z_1Z_2} I_{inj1} - \frac{g_{m1}Z_2}{1 - g_{m1}g_{m2}Z_1Z_2} I_{inj2}. \tag{3.10a}$$

$$I_{o2} = \frac{g_{m1}g_{m2}Z_1Z_2}{1 - g_{m1}g_{m2}Z_1Z_2} I_{inj2} - \frac{g_{m2}Z_1}{1 - g_{m1}g_{m2}Z_1Z_2} I_{inj1}. \tag{3.10b}$$

Define

$$H_{o1} = -g_{m2}Z_1, \tag{3.11a}$$

$$H_{o2} = g_{m1}Z_2, \tag{3.11b}$$

$$H_o = H_{o1}H_{o2}. \tag{3.11c}$$

Note that at the resonant frequency of the RLC tanks, both H_{o1} and H_{o2} are real as

$$Z_1(\omega_o), Z_2(\omega_o) = R. \tag{3.12}$$

Equation (3.10) can be written as

$$I_{o1} = -\frac{H_o}{1 + H_o} I_{inj1} - \frac{H_{o2}}{1 + H_o} I_{inj2}, \tag{3.13a}$$

$$I_{o2} = -\frac{H_o}{1 + H_o} I_{inj2} + \frac{H_{o1}}{1 + H_o} I_{inj1}. \tag{3.13b}$$

Equation (3.13) can be represented by the linear systems shown in Fig. 3.3. The block of H_{o1} is highlighted to emphasize that its transfer function is negative. It is seen from Fig. 3.3 that although both I_{inj1} and I_{inj2} are negated prior to their entry

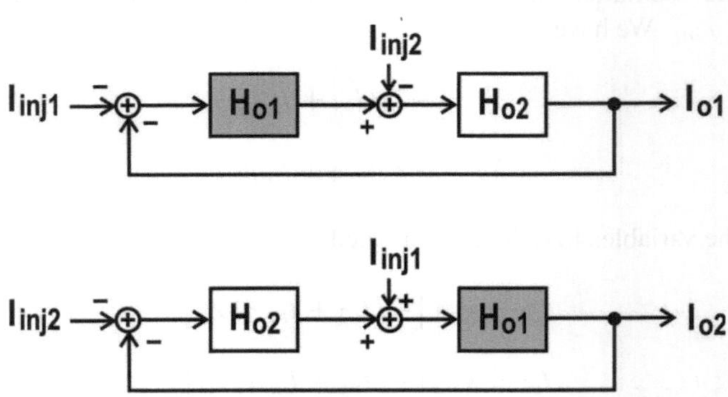

Fig. 3.3 Block diagram of the linear systems representing injection-locked harmonic oscillator of Fig. 3.1 in the lock state

to the system, since the transfer function from I_{inj2} to I_{o1} is positive while that from I_{inj1} to I_{o1} is negative, and I_{inj1} and I_{inj2} are out of phase, the two injected currents actually affect the output current differentially. The same holds for I_{o2} as well

When the injection signals are absent, the oscillator will oscillate at the self-resonant frequency ω_o of the resonators and Barkhausen criteria

$$1 + H_o(\omega_o) = 0 \tag{3.14}$$

will hold. When the injection signals i_{inj1} and i_{inj2} are present, the oscillator will be locked to the injection signal and the frequency of the oscillator will shift from ω_o to $\omega_{inj} = \omega_o + \Delta\omega$ with $\Delta\omega_o \ll \omega_o$ typically. Equation (3.13) can be written as

$$I_{o1} = -\frac{H_o(\omega_o + \Delta\omega)}{1 + H_o(\omega_o + \Delta\omega)} I_{inj1} - \frac{1}{H_{o1}(\omega_o + \Delta\omega)} \frac{H_o(\omega_o + \Delta\omega)}{1 + H_o(\omega_o + \Delta\omega)} I_{inj2}. \tag{3.15a}$$

$$I_{o2} = -\frac{H_o(\omega_o + \Delta\omega)}{1 + H_o(\omega_o + \Delta\omega)} I_{inj2} + \frac{1}{H_{o2}(\omega_o + \Delta\omega)} \frac{H_o(\omega_o + \Delta\omega)}{1 + H_o(\omega_o + \Delta\omega)} I_{inj1}. \tag{3.15b}$$

$H_o(\omega_o + \Delta\omega)$ can be approximated with its first-order Taylor series expansion at ω_o, as illustrated graphically in Fig. 3.4

$$H_o(\omega_o + \Delta\omega) \approx H_o(\omega_o) + \left(\frac{dH_o}{d\omega}\right)_{\omega_o} \Delta\omega. \tag{3.16}$$

we have

$$1 + H_o(\omega_o + \Delta\omega) \approx 1 + H_o(\omega_o) + \left(\frac{dH_o}{d\omega}\right)_{\omega_o} \Delta\omega. \tag{3.17}$$

Making use of (3.14), we arrive at

$$1 + H_o(\omega_o + \Delta\omega) = \left(\frac{dH_o}{d\omega}\right)_{\omega_o} \Delta\omega. \tag{3.18}$$

As a result,

$$\frac{H_o(\omega_o + \Delta\omega)}{1 + H_o(\omega_o + \Delta\omega)} \approx \frac{-1}{\left(\frac{dH_o}{d\omega}\right)_{\omega_o} \Delta\omega}. \tag{3.19}$$

Note that we have utilized

$$H_o(\omega_o + \Delta\omega) \approx H_o(\omega_o) = -1 \tag{3.20}$$

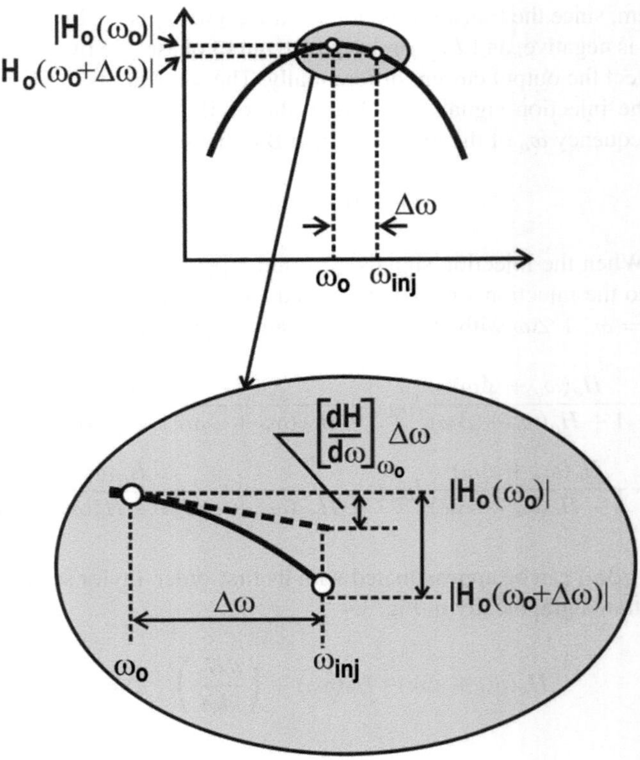

Fig. 3.4 Impact of injection signal on open-loop gain H_o of the oscillator of Fig. 3.1

in the simplification of (3.19) as

$$\left[\frac{dH_o}{d\omega}\right]_{\omega_o} \Delta\omega \ll H_o(\omega_o). \tag{3.21}$$

We arrive from (3.15)

$$I_{o1} \approx \frac{I_{inj1}}{\left(\frac{dH_o}{d\omega}\right)_{\omega_o} \Delta\omega} + \frac{1}{H_{o1}} \frac{I_{inj2}}{\left(\frac{dH_o}{d\omega}\right)_{\omega_o} \Delta\omega}, \tag{3.22a}$$

$$I_{o1}^* \approx \frac{I_{inj1}^*}{\left(\frac{dH_o^*}{d\omega}\right)_{\omega_o} \Delta\omega} + \frac{1}{H_{o1}^*} \frac{I_{inj2}^*}{\left(\frac{dH_o^*}{d\omega}\right)_{\omega_o} \Delta\omega}, \tag{3.22b}$$

$$I_{o2} \approx \frac{I_{inj2}}{\left(\frac{dH_o}{d\omega}\right)_{\omega_o} \Delta\omega} - \frac{1}{H_{o2}} \frac{I_{inj1}}{\left(\frac{dH_o}{d\omega}\right)_{\omega_o} \Delta\omega}, \tag{3.22c}$$

$$I_{o2}^{*} \approx \frac{I_{inj2}^{*}}{\left(\frac{dH_o^*}{d\omega}\right)_{\omega_o} \Delta\omega} - \frac{1}{H_{o2}^{*}} \frac{I_{inj1}^{*}}{\left(\frac{dH_o^*}{d\omega}\right)_{\omega_o} \Delta\omega}, \tag{3.22d}$$

where superscript * denotes complex conjugate operation. It follows from (3.22) that

$$I_{o1} I_{o1}^{*} \approx \frac{1}{\left(\frac{dH_o}{d\omega}\frac{dH_o^*}{d\omega}\right)_{\omega_o} (\Delta\omega)^2} \left(|I_{inj1}|^2 + \frac{|I_{inj2}|^2}{|H_{o1}|^2} + \frac{I_{inj1} I_{inj2}^{*}}{H_{o1}^{*}} + \frac{I_{inj1}^{*} I_{inj2}}{H_{o1}} \right), \tag{3.23a}$$

$$I_{o2} I_{o2}^{*} \approx \frac{1}{\left(\frac{dH_o}{d\omega}\frac{dH_o^*}{d\omega}\right)_{\omega_o} (\Delta\omega)^2} \left(|I_{inj2}|^2 + \frac{|I_{inj1}|^2}{|H_{o2}|^2} - \frac{I_{inj2} I_{inj1}^{*}}{H_{o2}^{*}} - \frac{I_{inj2}^{*} I_{inj1}}{H_{o2}} \right). \tag{3.23b}$$

Let

$$H_o = A e^{j\phi}, \tag{3.24}$$

where A and ϕ are the magnitude and phase of H_o, respectively. It follows that

$$H_o * = A e^{-j\phi}. \tag{3.25}$$

Because

$$\frac{dH_o}{d\omega} = e^{j\phi}\left(\frac{dA}{d\omega} + jA\frac{d\phi}{d\omega}\right), \tag{3.26a}$$

$$\frac{dH_o^*}{d\omega} = e^{-j\phi}\left(\frac{dA}{d\omega} - jA\frac{d\phi}{d\omega}\right), \tag{3.26b}$$

we have,

$$\frac{dH_o}{d\omega}\frac{dH_o^*}{d\omega} = \left(\frac{dA}{d\omega}\right)^2 + A^2\left(\frac{d\phi}{d\omega}\right)^2. \tag{3.27}$$

Evaluating (3.27) at $\omega = \omega_o$ and noting that

$$A^2 = 1 \tag{3.28}$$

as

$$1 + H_o(\omega_o) = 1 + A e^{j\phi} = 0, \tag{3.29}$$

we arrive at

$$\left(\frac{dH_o}{d\omega}\frac{dH_o^*}{d\omega}\right)_{\omega_o}=\left(\frac{dA}{d\omega}\right)_{\omega_o}^2+\left(\frac{d\phi}{d\omega}\right)_{\omega_o}^2. \tag{3.30}$$

Recall that the quality factor of harmonic oscillators was introduced in Chap. 2 and is repeated here for convenience

$$Q(\omega_o)=\frac{\omega_o}{2}\sqrt{\left(\frac{dA}{d\omega}\right)_{\omega_o}^2+\left(\frac{d\phi}{d\omega}\right)_{\omega_o}^2}. \tag{3.31}$$

Making use of (3.31), we arrive from (3.30)

$$\left(\frac{dH_o}{d\omega}\frac{dH_o^*}{d\omega}\right)_{\omega_o}=\left[\frac{2Q(\omega_o)}{\omega_o}\right]^2. \tag{3.32}$$

Substituting (3.32) into (3.23) yields

$$|I_{o1}|^2=\left[\frac{\omega_o}{2Q(\omega_o)}\right]^2\frac{1}{(\Delta\omega)^2}\left(|I_{inj1}|^2+\frac{|I_{inj2}|^2}{|H_{o1}|^2}+\frac{I_{inj1}I_{inj2}^*}{H_{o1}^*}+\frac{I_{inj1}^*I_{inj2}}{H_{o1}}\right), \tag{3.33a}$$

$$|I_{o2}|^2=\left[\frac{\omega_o}{2Q(\omega_o)}\right]^2\frac{1}{(\Delta\omega)^2}\left(|I_{inj2}|^2+\frac{|I_{inj1}|^2}{|H_{o2}|^2}-\frac{I_{inj2}I_{inj1}^*}{H_{o2}^*}-\frac{I_{inj2}^*I_{inj1}}{H_{o2}}\right). \tag{3.33b}$$

Solve for $\Delta\omega$

$$|\Delta\omega|=\frac{\omega_o}{2Q(\omega_o)}\sqrt{\left|\frac{I_{inj1}}{I_{o1}}\right|^2+\frac{1}{|H_{o1}|^2}\left|\frac{I_{inj1}}{I_{o1}}\right|^2+\frac{1}{H_{o1}^*}\frac{I_{inj1}I_{inj2}^*}{|I_{o1}|^2}+\frac{1}{H_{o1}}\frac{I_{inj1}^*I_{inj2}}{|I_{o1}|^2}}, \tag{3.34a}$$

$$|\Delta\omega|=\frac{\omega_o}{2Q(\omega_o)}\sqrt{\left|\frac{I_{inj2}}{I_{o2}}\right|^2+\frac{1}{|H_{o2}|^2}\left|\frac{I_{inj2}}{I_{o2}}\right|^2-\frac{1}{H_{o2}^*}\frac{I_{inj2}I_{inj1}^*}{|I_{o2}|^2}-\frac{1}{H_{o2}}\frac{I_{inj2}^*I_{inj1}}{|I_{o2}|^2}}. \tag{3.34b}$$

Equation (3.34) quantifies the amount of the deviation of the frequency of the oscillator from its natural frequency ω_o caused by the injection signals and is therefore the lock range of the oscillator for given injection strength. Note that

H_{o1} and H_{o2} are evaluated at ω_{inj}. It is seen that the lock range of the oscillator is inversely proportional to the quality factor of the oscillator at ω_o.

The dependence of the lock range on the strength of the injection signals is rather a bit complicated and deserves a special attention. Let us examine it in detail. The first two terms on the right-hand side of (3.34) quantify the individual contribution of I_{inj1} and I_{inj2} to the lock range of the oscillator. It is seen that due to the different injection locations of i_{inj1} and i_{inj2}, i_{inj1} contributes directly to the lock range while the contribution of i_{inj2} to the lock range is scaled down by $|H_{o1}|$. As to be seen shortly that since $|H_{o1}| = 1$, both injection currents contribute to the overall lock range equally.

If only one injection is present, for example, $i_{inj1} = i_{inj}$ and $i_{inj2} = 0$, Eq. (3.34) will be simplified to

$$|\Delta\omega| = \frac{\omega_o}{2Q}\left|\frac{I_{inj}}{I_{o1}}\right|. \tag{3.35}$$

Equation (3.34) is the same as the lock range of injection-locked harmonic oscillators with a single injection obtained in Chap. 2.

The last two terms on the right-hand side of (3.34) signify the joint contribution of I_{inj1} and I_{inj2}, which depends upon the phase relation between i_{inj1} and i_{inj2}. To analyze this contribution to the lock range, we rewrite (3.9) here for convenience

$$I_{o1} = g_{m1}\left[-\left(I_{o2} + I_{inj2}\right) Z_2\right]. \tag{3.36a}$$

$$I_{o2} = g_{m2}\left[-\left(I_{o1} + I_{inj1}\right) Z_1\right]. \tag{3.36b}$$

Utilizing (3.11a), we arrive from (3.36)

$$I_{o1} = -H_{o2}I_{o2} - H_{o2}I_{inj2}, \tag{3.37a}$$

$$I_{o2} = H_{o1}I_{o1} + H_{o1}I_{inj1}. \tag{3.37b}$$

The phase difference between v_1 and v_2 is 180° regardless whether the injection signals are present or not. Further since v_1 and v_2 have the same amplitude, their phasors at ω_o can be written as

$$V_1 = V_o e^{j\phi_{inj1}}, \tag{3.38a}$$

$$V_2 = V_o e^{j(\phi_{inj1}+\pi)} = -V_o e^{j\phi_{inj1}}. \tag{3.38b}$$

Since

$$I_{o1} = g_{m1}V_2, \tag{3.39a}$$

$$I_{o2} = g_{m2}V_1, \tag{3.39b}$$

and $g_{m1} = g_{m2} = g_m$, we have from (3.38) and (3.39)

$$I_{o1} = -g_m V_o e^{j\phi_{inj1}}, \tag{3.40a}$$

$$I_{o2} = g_m V_o e^{j\phi_{inj1}}. \tag{3.40b}$$

When the injection signals are absent, we will have from (3.37)

$$I_{o1} = -H_{o2} I_{o2}, \tag{3.41a}$$

$$I_{o2} = H_{o1} I_{o1}. \tag{3.41b}$$

As a result,

$$H_{o1} = -1, \tag{3.42a}$$

$$H_{o2} = 1. \tag{3.42b}$$

Equation (3.42) is valid at ω_o. Since $\Delta\omega \ll \omega_o$, we have

$$H_{o1}(\omega_o + \Delta\omega) \approx H_{o1}(\omega_o), \tag{3.43a}$$

$$H_{o2}(\omega_o + \Delta\omega) \approx H_{o2}(\omega_o). \tag{3.43b}$$

Let the two injection currents be given by

$$I_{inj1} = I_{inj} e^{j\phi_{inj1}}, \tag{3.44a}$$

$$I_{inj2} = I_{inj} e^{j\phi_{inj2}}. \tag{3.44b}$$

and

$$|I_{o1}| = |I_{o2}| = I_o. \tag{3.45}$$

The two expressions of (3.34) yield the same result

$$|\Delta\omega| = \frac{\omega_o}{2Q(\omega_o)} \left| \frac{I_{inj}}{I_o} \right| \sqrt{2 - \left(e^{j\Delta\phi_{inj}} + e^{-j\Delta\phi_{inj}} \right)}, \tag{3.46}$$

where

$$\Delta\phi_{inj} = \phi_{inj1} - \phi_{inj2}. \tag{3.47}$$

It is seen that the term preceding the square-root operator in (3.46) is the lock range of the oscillator with a single injection. We denote it by $|\Delta\omega|_s$

$$|\Delta\omega|_s = \frac{\omega_o}{2Q(\omega_o)}\left|\frac{I_{inj}}{I_o}\right|. \tag{3.48}$$

The square-root term in (3.46) quantifies the contribution of dual injections to the lock range of the oscillator. The lock range of the oscillator with dual current injections, denoted by $|\Delta\omega|_d$, is related to that with a single injection by

$$|\Delta\omega|_d = S_{inj}|\Delta\omega|_s, \tag{3.49}$$

where S_{inj} is given by

$$S_{inj} = \sqrt{2 - \left(e^{j\Delta\phi_{inj}} + e^{-j\Delta\phi_{inj}}\right)}. \tag{3.50}$$

S_{inj} quantifies the impact of injection signaling, specifically, the phase difference of the two injection signals on the lock range of the oscillator. It is seen in Fig. 3.5 that when $\Delta\phi_{inj} = 0$, since $e^{j0} + e^{-j0} = 2$, we will have $S_{inj} = 0$. As a result, $|\Delta\omega|_d = 0$. This result is as expected. The reason for this is as follows: It is seen from Fig. 3.3 that since i_{inj1} and i_{inj2} are in phase in this case while the transfer function from I_{inj1} to I_{o1} is positive and that from I_{inj2} to I_{o1} is negative (note that H_{o1} is negative while H_{o2} is positive), if I_{inj1} causes the frequency of I_{o1} to deviate

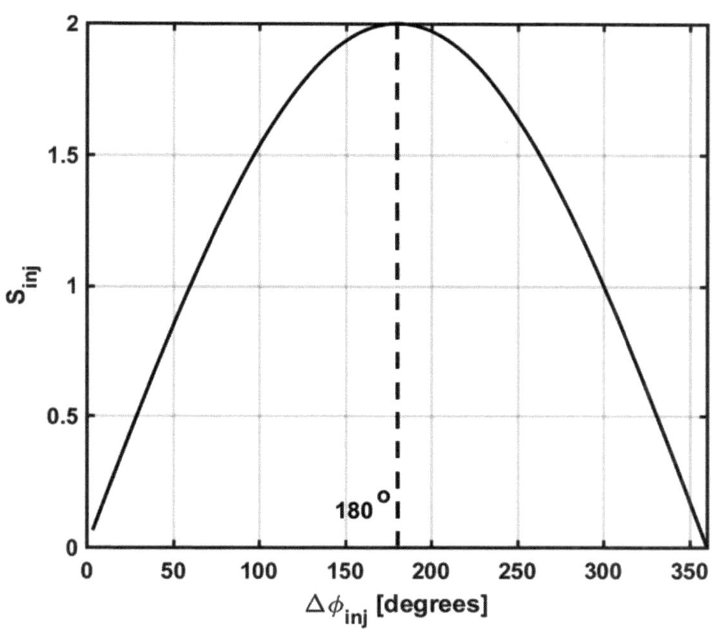

Fig. 3.5 Dependence of the lock range of the harmonic oscillator in Fig. 3.1 on the phase difference between the injection signals

from ω_o by $\Delta\omega$ in order to satisfy Barkhausen criteria, I_{inj2} will cause the frequency of I_{o1} to deviate from ω_o by $-\Delta\omega$ in order to satisfy Barkhausen criteria. Note that the magnitude of the transfer function from I_{inj1} to I_{o1} and that from I_{inj2} to I_{o1} are the same. As the system is linear, using the principle of superposition, we obtain that the net frequency variation induced by I_{inj1} and I_{inj2} in I_{o1} is zero. Similarly one can show that the net frequency variation induced by I_{inj1} and I_{inj2} in I_{o2} is also zero.

When $\Delta\phi = \pi$, since $e^{j\pi} + e^{-j\pi} = -2$, we will have $S_{inj} = 2$. The lock range in this case is maximized and is twice that with a single injection

$$|\Delta\omega|_d = 2|\Delta\omega|_s. \tag{3.51}$$

The reason for this is as follows: It is seen from Fig. 3.3 that since i_{inj1} and i_{inj2} are out of phase in this case, if I_{inj1} causes the frequency of I_{o1} to deviate from ω_o by $\Delta\omega$ in order to satisfy Barkhausen criteria, I_{inj2} will also cause the frequency of I_{o1} to deviate from ω_o by $\Delta\omega$ in order to satisfy Barkhausen criteria. The total frequency variation induced by I_{inj1} and I_{inj2} in I_{o1} in order to satisfy Barkhausen criteria is therefore $2\Delta\omega$. Similarly one can show that the net frequency variation induced by I_{inj1} and I_{inj2} in I_{o2} is also $2\Delta\omega$.

Figure 3.5 plots the dependence of S_{inj} on $\Delta\phi_{inj}$. The lock range will be minimized when the difference between the phase of I_{inj1} and that of I_{inj2} is zero and maximized when it is 180°.

As an example, consider the LC oscillator shown in Fig. 3.6 designed in an IBM 130 nm 1.2 V technology with free-running frequency 2.4 GHz [105]. The quality factor of the LC tank is set to 1.7. The phase of injection v_{inj1} is set to 0 while that of v_{inj2} is varied from 0° to 270° so as to allow us to explore the dependence of the lock range of the oscillator on the phase difference between the two injection signals. The lock range of the oscillator is tabulated in Table 3.1. For the purpose of comparison, the same LC oscillator with only v_{inj1} present, i.e., a single injection is

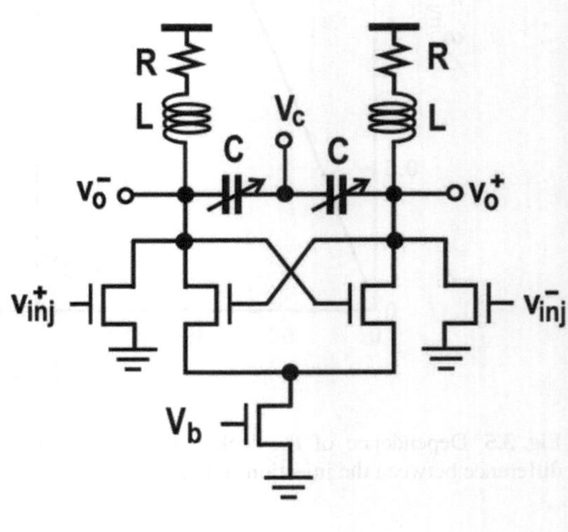

Fig. 3.6 Schematic of LC oscillator with differential injections

Table 3.1 Dependence of the lock range of the oscillator in Fig. 3.6 on the phase difference between the injection signals

$\Delta\phi$	$\Delta\omega_d$	$\Delta\omega_s$	$\Delta\omega_d/\Delta\omega_s$
0°	0	59 MHz	0
45°	45 MHz	59 MHz	0.76
90°	85 MHz	59 MHz	1.44
135°	107 MHz	59 MHz	1.81
180°	122 MHz	59 MHz	2.07
270°	85 MHz	59 MHz	1.44

also analyzed and the lock range of the oscillator is also tabulated in Table 3.1. It is seen that the lock range of the oscillator with two injection signals varies with the phase difference between them. It will peak when the phase difference is 180° and is twice that with only one injection signal. The lock range will be minimized when the phase difference between the injection signals is 0.

3.3 Inductor Series-Peaking

It was shown in Chap. 2 that ILFDs are popular in realizing the pre-scalar of frequency synthesizers, accredited to their ability to operate at a high frequency and low power consumption. Figure 2.9 shows that an ILFD can be represented by a nonlinear system composing of a nonlinear block $y = f(x)$ functioning as a mixer that mixes the input signal of the ILFD at ω_{inj} with the output of the ILFD at ω_o and a high-Q band-pass filter whose passband center frequency is set to ω_o. Both blocks reside in the forward path of the system. The functionality of the mixer is to generate the desired frequency component that falls into the passband of the resonator. For example, for a divide-by-2 ILFD, the input is at $2\omega_o$ and the desired frequency is at ω_o. For a divide-by-3 ILFD, the input is at $3\omega_o$ and the desired frequency is at ω_o. The resonator allows the frequency components that fall into its passband to pass through and reach the output of the oscillator while suppressing out all other frequency components generated in the mixing operation.

Due to the constraints of high operation frequencies and low power consumption imposed by most wireless applications, the nonlinear block of ILFDs for generating the desired frequency component is typically implemented using a single transistor. The generation of the desired frequency component is realized by utilizing the nonlinear characteristics of MOS transistors. Both gate-pumped [19] and drain-pumped single-transistor mixers [7, 10], as shown in Fig. 3.7, are used. The former utilize the nonlinear characteristics of $i_{DS} \sim v_{GS}$ relation of MOS transistors with v_{GS} just above the threshold voltage of transistors whereas the latter takes the advantage of the harsh nonlinear characteristics of $i_{DS} \sim v_{DS}$ relation of MOS transistors with v_{DS} just beyond the saturation voltage V_{sat} or the pinch-off point of MOS transistors.

In a drain-pumped mixer, the input and output of the mixer are applied to the gate and drain of the transistor, respectively, so as to maximize the conversion gain.

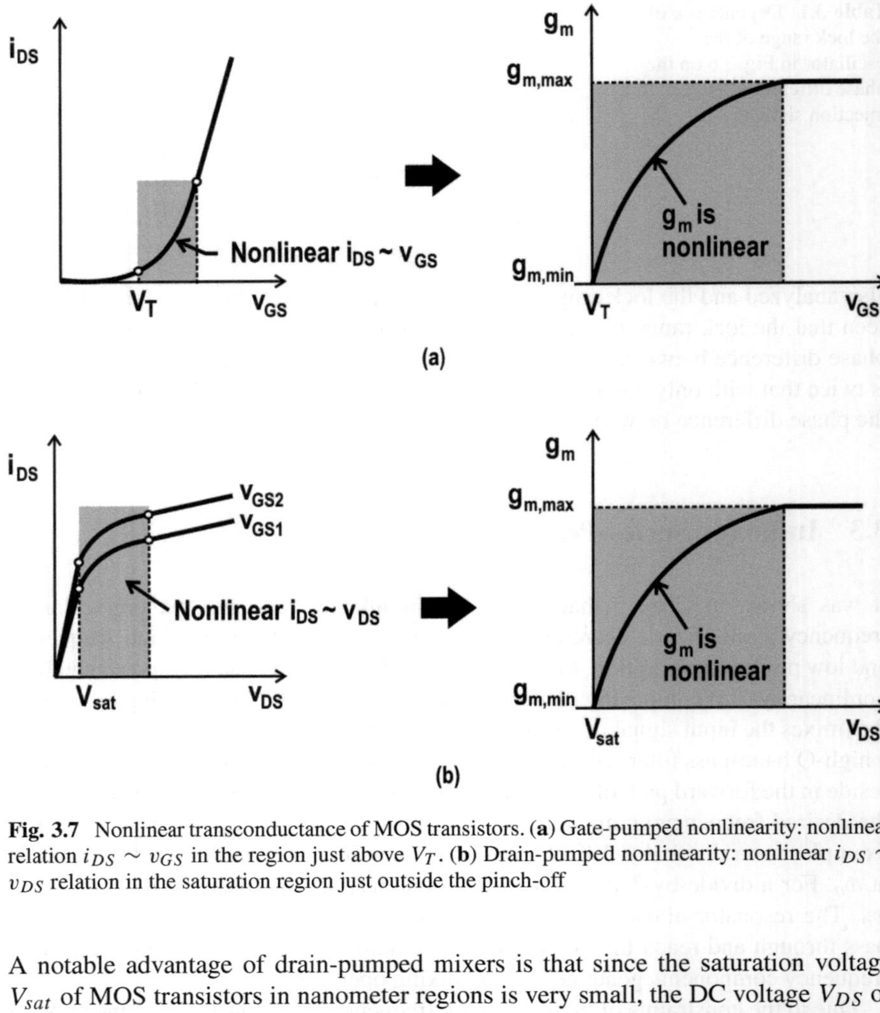

Fig. 3.7 Nonlinear transconductance of MOS transistors. (**a**) Gate-pumped nonlinearity: nonlinear relation $i_{DS} \sim v_{GS}$ in the region just above V_T. (**b**) Drain-pumped nonlinearity: nonlinear $i_{DS} \sim v_{DS}$ relation in the saturation region just outside the pinch-off

A notable advantage of drain-pumped mixers is that since the saturation voltage V_{sat} of MOS transistors in nanometer regions is very small, the DC voltage V_{DS} of the transistor can be made near zero [19], making drain-pumped mixers particularly attractive for low-voltage applications.

Figure 3.8 shows the simplified schematic of a drain-pumped configured ILFD. The resonators are formed by inductors L_s and the capacitances encountered at the drain and source of M4, which consists of the capacitances of M2, M3, M4, and those of the spirals. Transistors M1–M3 form a negative resistor to cancel out the parasitic resistance of the resonator arising from both the series resistance of the spirals and the output resistance of the transistors. The injection signal is applied to the gate of M4. M4 functions as a drain-pumped mixer that mixes v_{in} with $v_{DS4} = v_{o1} - v_{o2}$. It was shown in Fig. 2.26 of Chap. 2 that the main nonlinearity of drain-pumped mixers is the nonlinear transconductance g_m when the transistor operates in the saturation region just outside the pinch-off [7, 10, 94, 96]. Since $g_m = \frac{\partial i_{DS}}{\partial v_{GS}}$ and

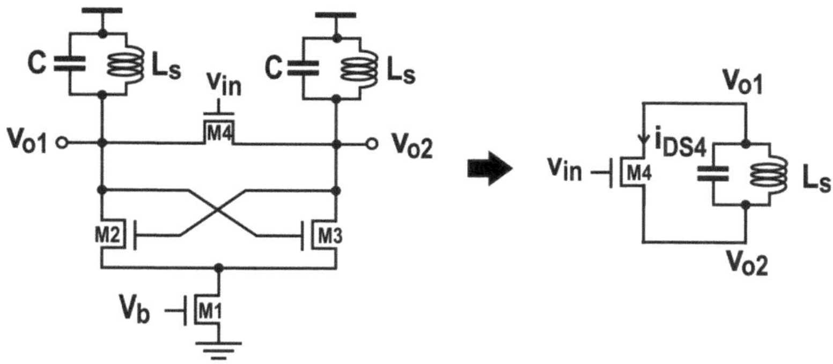

Fig. 3.8 Drain-pumped divide-by-2 injection-locked frequency dividers

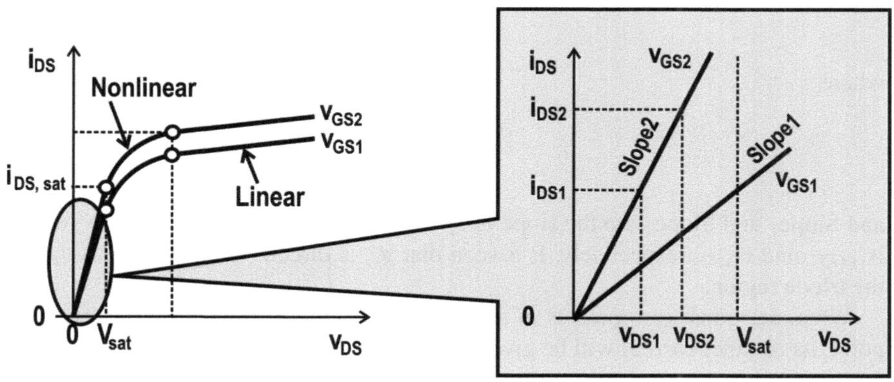

Fig. 3.9 Linear transconductance of MOS transistors when operating in the triode region

i_{DS} is a function of both v_{GS} and v_{DS}, i.e., $i_{DS} = f(v_{GS}, v_{DS})$, the channel current is modulated by both the gate voltage and drain-source voltage of the transistor in a nonlinear fashion. g_m varies with v_{DS} linearly in the triode in Fig. 3.9 and nonlinearly near and outside the pinch-off, as illustrated graphically in Fig. 3.10.

When the transistor operates in the triode region, we have the transconductance at v_{DS1} and v_{DS2}

$$g_{m1} = \frac{v_{DS1}\left(\text{slope}_2 - \text{slope}_1\right)}{V_{GS2} - V_{GS1}} = av_{DS1}, \tag{3.52}$$

$$g_{m2} = \frac{v_{DS2}\left(\text{slope}_2 - \text{slope}_1\right)}{V_{GS2} - V_{GS1}} = av_{DS2}, \tag{3.53}$$

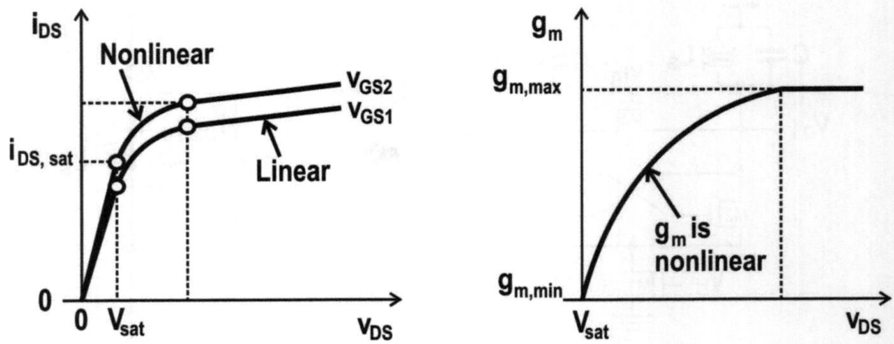

Fig. 3.10 Nonlinear transconductance of MOS transistors when operating in the saturation region just outside the pinch-off

where

$$a = \frac{\text{slope}_2 - \text{slope}_1}{V_{GS2} - V_{GS1}}, \tag{3.54}$$

and Slope$_1$ and Slope$_2$ are the slope of $i_{DS} \sim v_{DS}$ curve when gate-source voltage is v_{GS1} and v_{GS2}, respectively. It is seen that g_m is directly proportional to v_{DS} in the triode region.

When the transistor operates in a saturation region far away from the pinch-off point, its channel current will be given by

$$i_{DS} = i_{DS,sat}\left(1 + \lambda v_{DS}\right), \tag{3.55}$$

where $i_{DS,sat}$ is the channel current at the pinch-off and is given by

$$i_{DS,sat} = \frac{1}{2}\mu_n C_{ox}\left(\frac{W}{L}\right)\left(v_{GS} - V_T\right)^2, \tag{3.56}$$

where μ_n is the surface mobility of free electrons, C_{ox} is gate capacitance per unit area, and λ is channel-length-modulation coefficient. In this case, the transconductance is given by

$$\begin{aligned} g_m &= \frac{\partial i_{DS}}{\partial v_{GS}} \\ &= \mu_n C_{ox}\left(\frac{W}{L}\right)\left(v_{GS} - V_T\right)\left(1 + \lambda v_{DS}\right). \end{aligned} \tag{3.57}$$

Equation (3.57) shows that g_m is modulated by both v_{GS} and v_{DS} in a linear fashion. As λ is small, g_m is mainly modulated by v_{GS} when the transistor is in saturation.

When the transistor operates in the saturation region just outside the pinch-off, $i_{DS} \sim v_{DS}$ relation in this case will be nonlinear, as shown in Fig. 3.10. Most SPICE programs characterize $i_{DS} \sim v_{DS}$ relation in this region using a polynomial or a set of polynomials. Equation (3.55) can be written as

$$i_{DS} \approx i_{DS,sat}\left(1 + a_1 v_{DS} + a_2 v_{DS}^2 + a_3 v_{DS}^3\right). \tag{3.58}$$

The transconductance of the transistor is therefore given by

$$g_m = \mu_n C_{ox}\left(\frac{W}{L}\right)(v_{GS} - V_T)\left(1 + a_1 v_{DS} + a_2 v_{DS}^2 + a_3 v_{DS}^3\right). \tag{3.59}$$

Equation (3.59) shows that g_m is modulated by v_{GS} linearly and by v_{DS} nonlinearly.

If an injection signal is applied to the gate of transistor M4 that is biased in saturation just outside the pinch-off while the output of the oscillator is applied across the drain and source of the transistor, the channel current of the transistor will be modulated by both the injection signal and the output of the oscillator and will contain multiple frequency components. As transistors M1–M3 form a negative resistor to compensate for the resistive loss of the resonator, the channel current of M4 flows into the resonator and only the frequency components of the channel current of M4 that fall into the passband of the resonators appear at the output nodes of the oscillator while all other frequency components are suppressed by the resonators. This can be further illustrated using the AC circuit shown in Fig. 3.8. The channel current of M4 contains various frequency components including the one that falls into the passband of the resonator. Assume that the negative resistance completely compensates for the resistive loss of the resonator. Since the resonator is lossless and functions as an open-circuit at ω_o and a short-circuit at all other frequencies, it will short all frequency components other than ω_o.

It is seen from Fig. 3.10 and (3.59) that the larger the variation of v_{DS}, the more severe the impact of v_{DS} on the channel current subsequently the larger the frequency components generated by the drain-pumped mixing operation including the one whose frequency falls into the passband of the resonator. As detailed in Chap. 2 that increasing the amplitude of the signal whose frequency falls into the passband of the resonator gives rise to a larger frequency deviation from the free-running frequency of the oscillator subsequently a larger lock range. In order to increase the lock range of drain-pumped ILFDs, v_{DS} of the injection transistor needs to be increased. This can be achieved by placing LC tanks at both the source and drain of the transistor M4, as shown in Fig. 3.11 [14]. The self-resonant frequency of the added LC resonators should be tuned to the output frequency of the ILFD.

To analyze the added resonators, consider the sub-circuit shown in Fig. 3.11. Since v_{o1} and v_{o2} are sinusoids with a phase difference of 180°, the added LC resonators have the transfer function

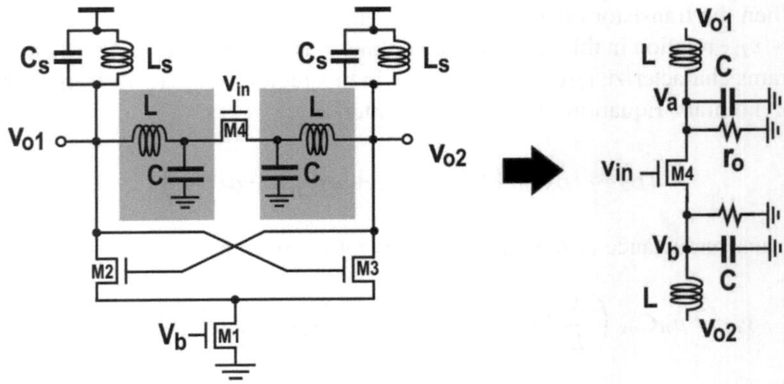

Fig. 3.11 Drain-pumped divide-by-2 injection-locked frequency dividers with series inductor peaking

$$\frac{V_a}{V_{o1}} = \frac{1}{LC}\frac{1}{s^2 + s\frac{1}{r_oC} + \frac{1}{LC}}, \quad (3.60a)$$

$$\frac{V_b}{V_{o2}} = \frac{1}{LC}\frac{1}{s^2 + s\frac{1}{r_oC} + \frac{1}{LC}}, \quad (3.60b)$$

where r_o is the output resistance of the transistor. Note that the transfer function is measured with the inputs at V_{o1}, V_{o2} and outputs V_a, V_b. It is seen from (3.60) that the gain provided by the added resonators is unity at low frequencies and peaks at the pole frequency $\omega_{LC} = \frac{1}{\sqrt{LC}}$. The value of L should be chosen in such a way that the gain provided by the resonators is maximized. Since v_{o1} and v_{o2} are out of phase while the two added LC resonators are identical, the amount of the phase shift provided by the two added resonators is the same. As a result, the voltage at the source and that at the drain are also out of phase, yielding the maximum v_{DS}.

Inductor series-peaking can also be used in divide-by-3 ILFDs. It was shown in Chap. 2 that divide-by-3 ILFDs achieve a frequency division of 3 by means of differential injection. To increase the lock range of divide-by-3 ILFDs without increasing the strength of the injection signal, the differential voltage generated by the injection signal and fed to the sources of cross-coupled NMOS transistors or the drains of cross-coupled PMOS transistors where frequency mixing between the injection signal and the output of the ILFD takes place needs to be maximized. We showed in Fig. 2.28 that this can be achieved by employing a pair of transformers with injection signals applied to the primary winding of the transformers and the sources of the mixing transistors connected to the secondary windings of the transformers. Alternatively, one can use a transformer with the primary and secondary windings of the transformer connected to the sources of the mixing transistors, as shown in Fig. 3.12 [34]. The resonator of the ILFD is formed by

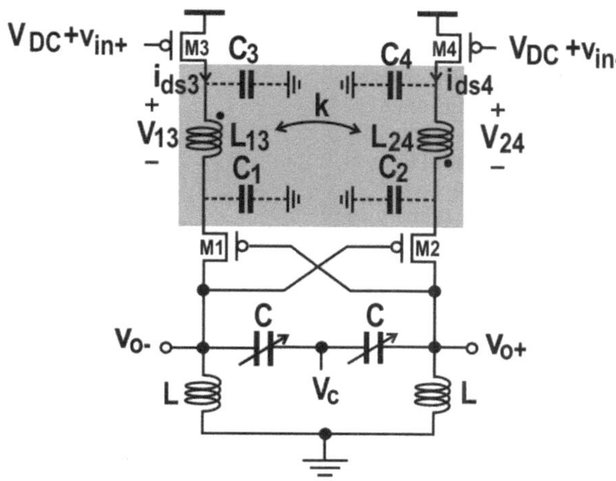

Fig. 3.12 Divide-by-3 ILFD with transformer series-peaking

inductors L and capacitors C that resonates at $\omega_o = 1/\sqrt{LC}$. The negative resistance provided by M1 and M2 compensates for the resistive loss of the spirals of both the inductors and the transformer. Note the polarity of the windings of the transformer and those of the inductors of the resonating tank due to the differential operation of the frequency divider. Transistors M3 and M4 operate in saturation and function as linear transconductors that generate differential currents $i_{ds3} = g_m v_{in+}$ and $i_{ds4} = g_m v_{in-}$ where $g_{m3}, g_{m4} = g_m$. Note since $v_{in}^+ = -v_{in}^-$, we have $i_{ds3} = -i_{ds4}$. The primary winding of the transformer resonates with the capacitors at the drain of M3 at $3\omega_o$. Similarly the secondary winding of the transformer resonates with the capacitor at the drain of M4 at $3\omega_o$. As a result, the drain of M3 and that of M4 see a large impedance. This is echoed with a large voltage swing at the drain of M3 and that of M4.

One might argue what is the advantage of using a transformer in the ILFD in Fig. 3.12 as compared with the divide-by-3 ILFD in Fig. 3.16 where an inductor is used to boost the injection signal prior to its mixing with the output of the ILFD? When an inductor is used, the differential injection will ensure that the center point of the inductor is an AC ground. As a result, the inductor can be split into two inductors of the same inductance, each forms a LC resonator with the capacitance connected to it. Both resonators resonate at $3\omega_o$ thereby eliminating the low-impedance path from the drain of the injection transistors to the ground at $3\omega_o$ and improving injection efficiency subsequently lock range. Note that when this occurs, the impedance seen by the drain of the injection transistors will be the source of the mixing transistors. The added inductor increases the impedance seen by the drain of the injection transistors by resonating out the low-impedance capacitive paths to the ground.

When the transformer is used, the close coupling of the transformer subsequently the large mutual inductor M will result in a large effective inductance $L_{13} + M$ at the primary winding and $L_{24} + M$ at the secondary winding where M is the

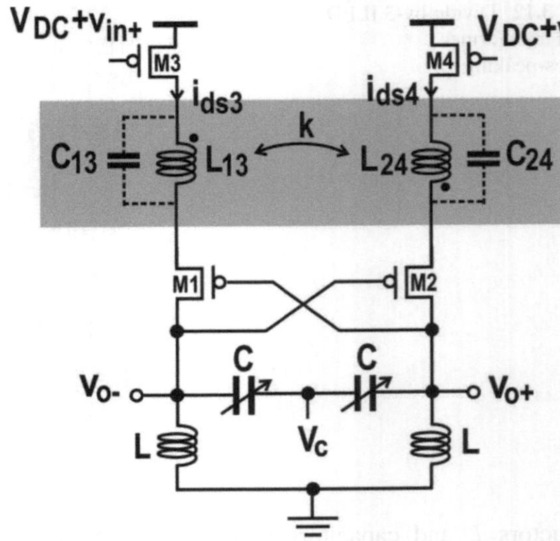

Fig. 3.13 Divide-by-3 ILFD with transformer series-peaking. The parasitic capacitances at the drain of M3–M4 and the source of M1–M2 are lumped to form C_{13} and C_{24} that are in parallel with the primary and secondary windings of the transformer

mutual inductance of the transformer with a small number of turns, thereby lowering the silicon area. Another key difference between the inductor shunt-peaking and transformer series-peaking is that the spirals of the transformer are placed between the drain of the injection transistors and the source of the mixing transistors. As C_1 and C_3 are in series, one can use a capacitor $C_{13} = C_1C_3/(C_1 + C_3)$ to replace C_1 and C_3. Note C_{13} is in parallel with L_{13}. The same holds for C_2 and C_4 as well, as shown in Fig. 3.13. When these LC tanks resonate at $3\omega_o$, they will behave as an open-circuit. As a result, the impedance seen by the drain of M3 and M4 will be maximized. A small differential injection signal can therefore generate a large voltage swing at the sources of M1 and M2, resulting in a large lock range.

3.4 Inductor Shunt-Peaking

As the frequency of a harmonic oscillator is set by the amount of the time required to charge and discharge its load capacitors, in order to maximize the impact of an injection signal on the frequency of the oscillator so as to yield the maximum lock range, the transfer function from the injection signal to the output of the oscillator measured at the frequency of the injection signal needs to be maximized. Figure 3.14 shows the schematic of a divide-by-2 ILFD with the injection signal applied to the gate of the tail current source transistor M1. The injection voltage is converted to current by M1: $i_{ds1} = g_{m1}v_{inj}$. i_{ds1} is steered to the output nodes by M2 and M3. The steered current discharges one of the output capacitors of the oscillator, for example, C_{o-}, while the other capacitor C_{o+} is charged by the V_{DD} via inductor L_s. Note that along with the discharge of C_{o-}, v_{o-} drops. This helps reducing the

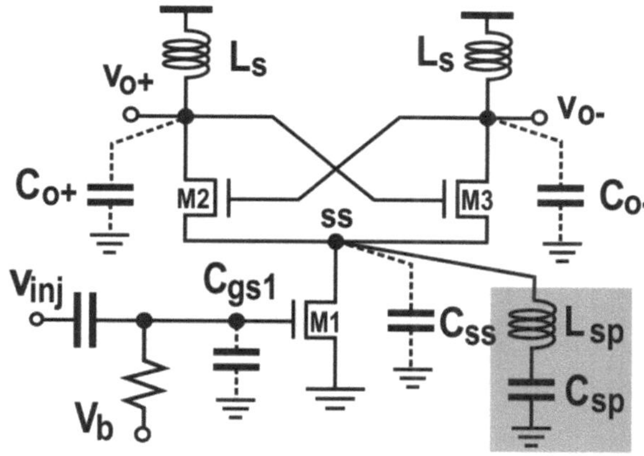

Fig. 3.14 Divide-by-2 ILFD with inductor shunt-peaking

channel current of M2 thereby improving the charging of C_{o+}. It becomes clear that injection speeds up both the discharging process of C_{o-} and the charging process of C_{o+}, thereby changing the frequency of the oscillator. The amount of the frequency shift that the injection signal generates clearly depends upon the strength of the injection signal.

The capacitance at node ss consists of capacitance contributed by M1, M2, and M3 and is therefore large. i_{ds1} must pass through node ss in order to reach M2 and M3 where the mixing of i_{ds1} and the output of the oscillator takes place. The large capacitance at node ss creates a low impedance path to the ground at high frequencies that shunts most of the injection signal to the ground. As a result, the impact of the injection signal on the charging and discharging of the output capacitors of the oscillator is reduced. This is echoed with a low injection efficiency subsequently a small lock range. It becomes evident that in order to maximize the effective injection signal, i.e., the signal that mixes with the output of the oscillator so as to maximize the lock range, the detrimental effect of C_{ss} must be minimized.

The loss of high-frequency signals at nodes where a large shunt capacitance exists is widely encountered in RF mixers. Figure 3.15 shows the schematic of a double-balanced mixer. The LC tanks formed by L_{IF} and C_{IF} at the immediate frequency (IF) ports of the mixer resonate at ω_{IF}. These LC tanks serve two distinct purposes: (a) Boost the IF signal and (b) filter out frequency components at frequencies other than ω_{IF}. Inductors L_{s1} and L_{s2} at the sources of M1 and M2 form source degeneration feedback to improve the linearity of $i_{DS1,2} \sim v_{RF+}, v_{RF-}$ relation of M1 and M2. The capacitor C_{ss34} at the source of M3 and M4 and C_{ss56} at the source of M5 and M6 are typically large simply due to the connection of three transistors to the nodes. They shunt a significant portion of the RF currents generated by M1 and M2 to the ground thereby lowering the conversion gain of the mixer. A technique commonly used in RF mixers to minimize this drawback is to use inductors $L_{s3} \sim L_{s6}$ at the sources of M3–M6 to resonate out C_{ss34} and C_{ss56} at the RF frequency such that i_{RF+} and i_{RF-} will reach M3–M6 where frequency mixing takes place without any attenuation.

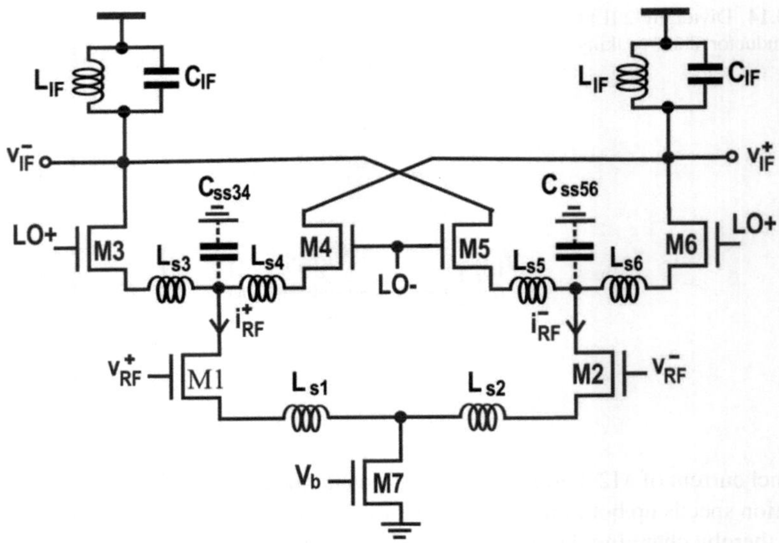

Fig. 3.15 Double-balanced RF mixer with an improved conversion gain

The preceding approach for improving the conversion gain of RF mixers was used in [91] to eliminate the detrimental effect of C_{ss} in the ILFD of Fig. 3.14 so as to ameliorate the lock range of the oscillator. Inductor L_{sp} is employed to resonate out C_{ss} at the frequency of the injection signal, i.e.,

$$\omega_{inj} = \frac{1}{\sqrt{L_{sp}C_{ss}}}. \tag{3.61}$$

Note that a DC isolation capacitor C_{sp} is needed to eliminate the direct current path from node ss to the ground. Since L_{sp} and C_{sp} will become a short-circuit at their resonant frequency

$$\omega_{sp} = \frac{1}{\sqrt{L_{sp}C_{sp}}}, \tag{3.62}$$

ω_{sp} should be made much higher than ω_{inj}.

Inductor shunt-peaking is also used in divide-by-3 ILFDs, as shown in Fig. 3.16. Transistors M2 and M3 operate in saturation and function as the transconductors that convert v_{inj}^+ and v_{inj}^- to i_{ds2} and i_{ds3}, respectively. The capacitance encountered at the drain of M2 and M3, denoted by C_{24} and C_{35}, respectively, is large due to the contribution of more than one transistor. They shunt a large portion of high-frequency injection currents i_{ds2} and i_{ds3} to the ground thereby lowering the lock range. To minimize this detrimental effect, shunt inductor L_{ss} is added between the drain of M2 and that of M3. The differential injection of the ILFD ensures that the

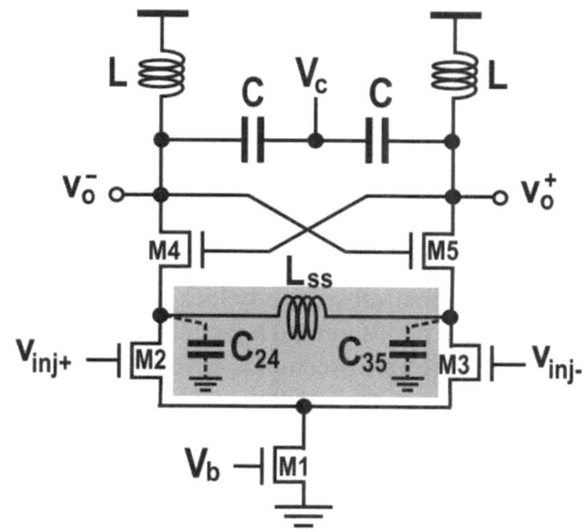

Fig. 3.16 Divide-by-3 ILFD with inductor shunt-peaking

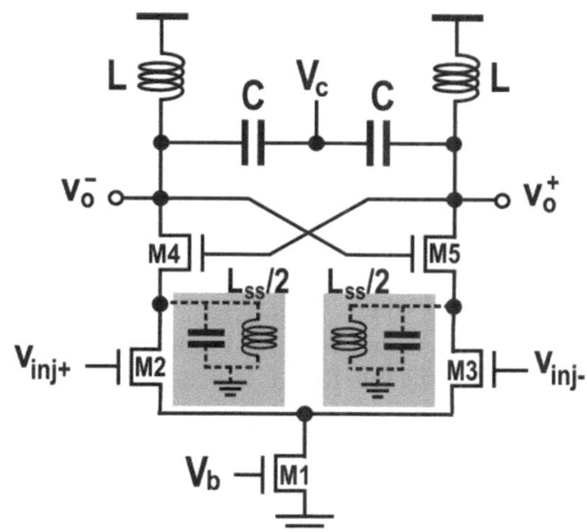

Fig. 3.17 Divide-by-3 ILFD with inductor shunt-peaking. L_{ss} can be split into two inductors of equal inductance $L_{ss}/2$, each forms a LC resonator with the capacitance connected to it

center point of the spiral of L_{ss} is an AC ground. As a result, L_{ss} can be split into two inductors of equal inductance $L_{ss}/2$, each forms a LC resonator with the capacitance connected to it, as shown in Fig. 3.17. The LC resonators resonate at $3\omega_o$ thereby eliminating the detrimental effect of C_{24} and C_{35} on the injection signal at $3\omega_o$.

3.5 Current-Reuse Direct Injection

It was shown earlier that a drain-pumped divide-by-2 ILFD offers better injection efficiency over a divide-by-2 ILFD using tail injection as the injection signal in the drain-pumped divide-by-2 ILFD is injected directly into the resonator, thereby minimizing the loss of the injection signal in the path from the injection node to the mixing devices where the desired frequency component is generated. This is known as direct injection. Figure 3.18 shows the schematic of a drain-pumped divide-by-2 ILFD. Note that a proper DC basing voltage is needed in order for M3 to operate in the saturation region just outside the pinch-off where $i_{DS} \sim v_{DS}$ nonlinearity is most dreadful. As the desired frequency component is generated by the drain-pumped nonlinearity of the injection transistor M3, in order to increase the effective injection signal subsequently lock range, the single-ended injection in Fig. 3.18 can be modified to differential injection as shown in Fig. 3.19 [8, 74]. Two cross-coupled transistor pairs are used to improve the transition edges subsequently the phase noise performance of the oscillator [25]. Another benefit of having both NMOS cross-coupled pair and PMOS cross-coupled pair is that each of them forms a negative resistor with resistance $R_N = -2/g_{m,n}$ provided by the NMOS pair and $R_P = -2/g_{m,p}$ provided by the PMOS pair where $g_{m,n}$ and $g_{m,p}$ are the transconductance of the NMOS and PMOS transistors, respectively. The overall negative resistance given by $R_N||R_P$ is smaller than R_N and R_P. Since the parasitic resistance of the spiral is typically small, g_m of the transistors of the negative resistor needs to be large. This is at the cost of both a high level of power consumption and a large silicon area. The use of two synthesized negative resistors relaxes this constraint. As the current of the PMOS cross-coupled pair is also used by the NMOS cross-coupled pair, this architecture is known as current-reuse direct injection.

The deployment of two injection transistors for drain-pumped mixing doubles injection efficiency. This is because since $i_{ds5} = f(v_{inj}^+, v_{DS})$ and $i_{ds6} = f(v_{inj}^-, v_{DS})$ where $v_{DS5}, v_{DS6} = v_{DS}$, i_{ds5} and i_{ds6} are the channel current of M5 and M6, respectively, and function f depicts $i_{DS} \sim v_{DS}$ nonlinear relation, the current flowing into the LC tank is given by $i_{LC} = i_{ds5}+i_{ds6}$. Since i_{ds5} and i_{ds6} are

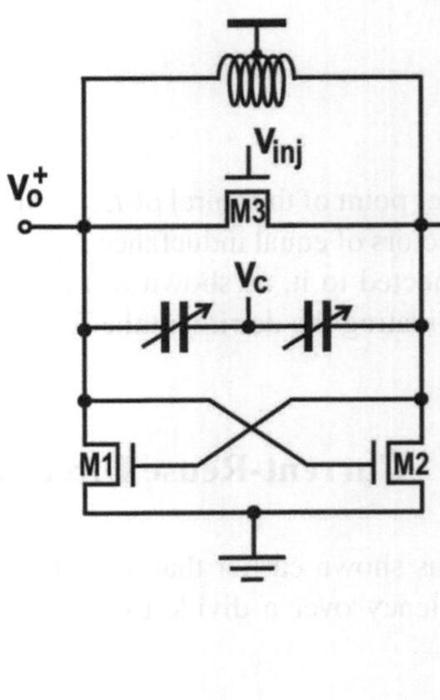

Fig. 3.18 Divide-by-2 drain-pumped ILFD

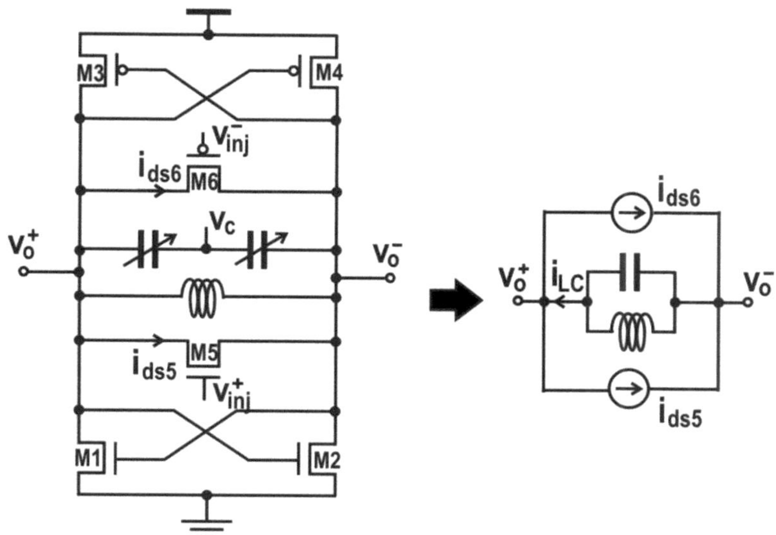

Fig. 3.19 Divide-by-2 drain-pumped ILFD with current reuse

modulated by the same $v_{DS} = v_o^+ - v_o^-$, the desired frequency component whose frequency falls into the passband of the LC resonator is doubled.

3.6 Quality Factor Reduction

It was shown earlier that inductor series-peaking increases lock range by boosting the voltage swing of v_{DS} of the transistor where frequency mixing between the injection signal and the output of ILFDs takes place without increasing the strength of the injection signal. The price paid for this is a large silicon area due to the use of two spiral inductors. Inductor shunt-peaking increases lock range by boosting the transfer function from the injection node to the frequency mixing node where the effective injection signal is generated without increasing the strength of the injection signal. The price paid for this again is a large silicon area due to the use of an extra spiral inductor. Since the lock range of harmonic oscillators is directly proportional to the strength of injection signals and inversely proportional to the quality factor of the resonator of the oscillators, a large lock range can also be obtained by lowering the quality factor of the oscillators without increasing the strength of the injection signal. In [11], a resistor R_s is added in series with the spiral inductor L_s of the resonators of oscillators to lower the quality factor of the resonators, as shown in Fig. 3.20. Note that the negative resistor formed by M1–M3 is designed to only compensate for the resistive loss caused by the series resistance of the spirals and

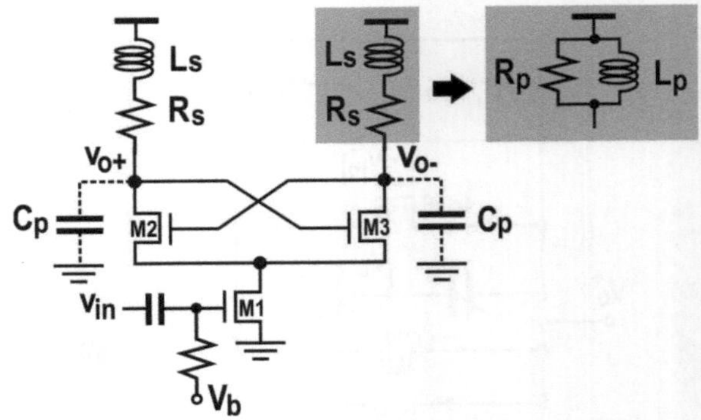

Fig. 3.20 Injection-locked frequency dividers with quality factor reduction

the output resistance of the transistors, the added series resistors is not and should not be compensated by the negative resistor.

To analyze the effect of the increased series resistance of the resonators, we utilize the relation between a series RL network consisting of L_s and R_s and a parallel RL network consisting of L_p and R_p studied in Chap. 2. The series RL network can be converted to the parallel RL network under the condition that the impedance looking into both networks is identical. This leads to

$$R_p = R_s(1 + Q_s^2) \tag{3.63}$$

and

$$L_p = L_s\left(1 + \frac{1}{Q_s^2}\right), \tag{3.64}$$

where

$$Q_s = \frac{\omega L_s}{R_s}. \tag{3.65}$$

It is evident that lowering Q_s will lower R_p. Since the gain of the inverting stage of the oscillator is maximized at the resonant frequency of the resonator and is given by

$$A_v \approx -g_{m2,3}(R_p||r_{o2,3}), \tag{3.66}$$

lowering Q_s will lower the gain of each of the two stages of the oscillator subsequently the loop gain of the oscillator given by

$$|H_o(j\omega_o)| = \left[g_{m2,3}(R_p||r_{o2,3})\right]^2. \tag{3.67}$$

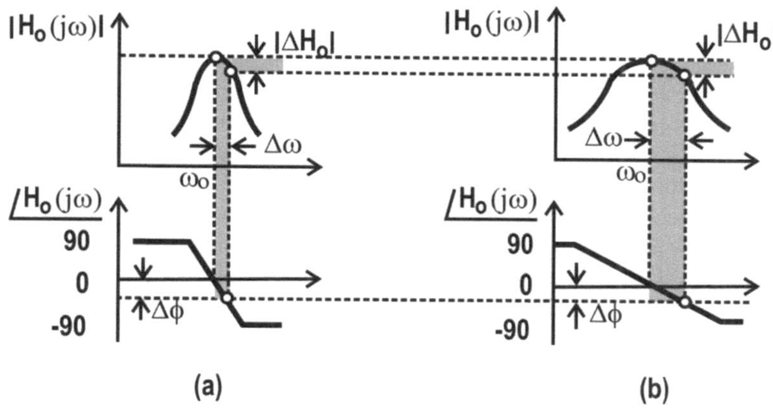

Fig. 3.21 Impedance of parallel RLC network with (**a**) high- and (**b**) low-quality factors

If the drop of the loop gain is overly large, Barkhausen magnitude criterion will be violated and oscillation will vanish. It is seen in Fig. 3.21 that when Q is lower, in order to have the same $|\Delta H_o|$ and $\Delta\phi$ induced by the injection signal so as to satisfy Barkhausen criteria, a larger frequency shift is needed.

Although lowering the quality factor via increasing R_s increases lock range, the upper bound of R_s is set by Barkhausen magnitude criterion. In other words, if the series resistance is overly large, a large drop of the impedance of the resonator will occur. As a result, Barkhausen criteria will not be satisfied and the oscillator will seize oscillation. Although one might argue that the increased series resistance can be offset by increasing the negative resistance formed by M1–M3 so as to not impact on the oscillation of the oscillator, our goal here is to lower the quality factor of the resonators with the negative resistance unchanged. Barkhausen magnitude criterion therefore sets the limit on how much we can lower the quality factor.

One question that we should ask ourselves is what is the price paid for a larger lock range by lowering the quality factor of the resonators. It is well understood that lowering the quality factor of harmonic oscillators will deteriorate the phase noise of the oscillators, as detailed in Chap. 2. Should we be concerned with the deterioration of the phase noise of oscillators once the quality factor is lowered? Fortunately although the phase noise of a free-running LC oscillator deteriorates once the quality factor of the oscillator is lowered as the PSD of the output noise of a free-running harmonic oscillator is inversely proportional to the square of the quality of the oscillator, if the oscillator is locked to an injection signal whose phase noise is better than that of the free-running oscillator, the feedback mechanism of injection-locking will ensure the phase wandering of the oscillator be confined to a small range such that the phase noise of the injection-locked oscillator is comparable to that of the injection signal when the frequency of the injection signal is close to the free-running frequency of the oscillator, as demonstrated both theoretically and numerically in Sect. 2.9 [11, 14, 18, 66, 91].

3.7 Resistor Feedback

It was shown earlier that resonant methods such as inductor series-peaking and inductor shunt-peaking boost the effective injection signal, i.e., the signal applied to the nonlinear core of ILFDs where the desired frequency component is generated subsequently lock range without increasing the strength of external injection signals. These approaches, however, are costly due to the need for area-greedy spiral inductors. Lowering the quality factor by adding resistors in series with the spirals of the resonators is also capable of increasing lock range without increasing injection signals. The value of the added resistor, however, is limited by Barkhausen criteria, which limits the amount of the lock range improvement that it can provide. In [35], a resistor feedback approach that feedbacks part of the output of the oscillator of ILFDs to the injection node was proposed to increase the overall injection signal subsequently lock range. This approach was extended to divide-by-3 ILFDs in [99], as shown in Fig. 3.22. Inductor L_2 and capacitors C form the resonator that resonates at ω_o. M5 and M6 form a negative resistor to compensate for the ohmic loss of the spiral. M3 and M4 also form a negative resistor at ω_o to further compensate for the resistive loss of the spiral at ω_o. Transistors M1 and M2 convert differential voltages v_{inj+} and v_{inj-} at $\omega_{inj} = 3\omega_o$ to channel currents i_{ds1} and i_{ds2}, respectively. The DC biasing voltage of M1 and M2 is the common-mode voltage of the outputs of the ILFD. Isolation capacitors C_{inj} are needed as they prevent the DC biasing voltage of M1 and M2 from loading the injection signal sources. M3 and M4 perform frequency mixing with the signals coming from their sources at $3\omega_o$ and their gates at ω_o primarily. Note that due to the finite quality factor of LC tank formed by L2 and the varactors, the frequency components of the drain voltage of M3 and M4 not only

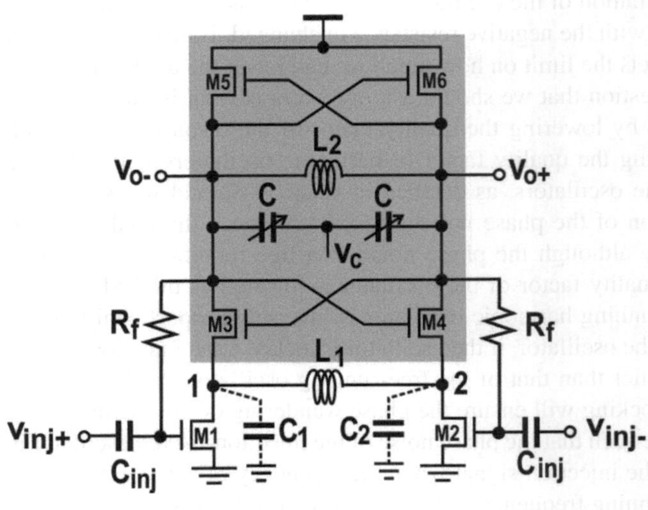

Fig. 3.22 Divide-by-3 ILFD with shunt resistor feedback

contains the desired frequency component at ω_o, but also contains other frequency components. The feedback paths provided by R_f allow frequency components at $3\omega_o$ be routed back to the gate of M1 and M2 and amplified by M1 and M2 so as to enhance $3\omega_o$ frequency components at the sources of M3 and M4.

The large value of C_1 and C_2 arising from the contribution from M1–M4 tends to shunt i_{ds1} and i_{ds2} to the ground, thereby lowering injection efficiency subsequently lock range. To eliminate this detrimental effect, inductor L_1 is added to resonate out C_1 and C_2 at $\omega_{inj} = 3\omega_o$, ensuring that i_{ds1} and i_{ds2} flow to the sources of M3 and M4, respectively with the minimum loss. Note that the differential injection ensures that the center node of the spiral of L_1 is an AC ground. As a result, L_1 can be split into two inductors of equal inductance $L_1/2$. One forms a resonator with C_1 and the other another resonator with C_2. Both resonates at

$$3\omega_o = \frac{1}{\sqrt{(L_1/2)C_{1,2}}}. \tag{3.68}$$

The LC tanks formed by L_1 and $C_{1,2}$ also behave as short circuits at frequencies other than ω_{inj}, thereby grounding nodes 1 and 2. This ensures the highlighted part of the ILFD resonates at ω_o such that the output of the ILFD is at ω_o.

The generation of the desired frequency component at ω_o is carried out by M3 and M4 whose source voltage is at $3\omega_o$ as all other frequency components of the source voltage are shunted to the ground by the LC tanks formed by L_1 and $C_{1,2}$ ideally. The drain voltage of M3 and M4 only contains frequency components at ω_o as all other frequency components of the drain voltage of M3 and M4 are shunted to the ground by the LC tanks formed by L_2 and C ideally. Other frequency components, however, do exist at the drain of M3 and M4 in reality due to the finite quality factor of the LC resonator. The same holds for the source voltage of M3 and M4 as well. The mixing operation performed by M3 and M4 with their source voltage at $3\omega_o$ and gate voltage at ω_o gives rise to multiple frequency components at the drain of M3 and M4 including those at $3\omega_o$. The frequency components of the drain voltage of M3 and M4 at ω_o are coupled back to the gate of M1 and M2 via R_f, thereby increasing the total injection voltage at the gate of M1 and M2. This in turn increases lock range.

3.8 Second Harmonic Extraction

It was shown in Chap. 2 that the divide-by-3 ILFD shown in Fig. 2.27 achieves a frequency division of 3 by mixing the source voltage of the mixing transistors at $3\omega_o$ and the gate voltage of the mixing transistors at ω_o in a nonlinear fashion so as to generate the desired frequency component at ω_o. Since the generation of the desired frequency component at ω_o is solely determined by the nonlinearity of the mixing transistors, only a fixed lock range can be obtained. For example, the lock range in [93] is 5.1% with 0 dBm injection strength while that in [33] is 9.2% with 0 dBm injection strength. Note that the latter employs a transformer between the injection signal and the transconductors to boost the output current of the transconductors prior to frequency mixing. As the frequency of the injection signal of a divide-by-3 ILDS is $3\omega_o$, a frequency division of 3 can be obtained if a mechanism capable of generating a frequency component at $2\omega_o$ prior to mixing with the injection signal exists and the strength of such a mechanism can be adjusted so as to increase lock range.

In [71], an extraction technique capable of extracting the second harmonic of the output of harmonic oscillators was proposed. Figure 3.23 shows how the second harmonic of the output of the oscillator is extracted. The output of the oscillator is dictated by the LC resonator formed by the shunt inductor L and the varactors C that resonates at ω_o. When v_o^+ is sufficiently high while v_o^- is sufficiently low, M1 will switch off while M2 will enter triode mode. Since in this case $v_{SD2} \approx 0$, $v_{ss1} \approx v_o^+$. Similarly, M4 will switch off while M3 will enter triode mode, forcing $v_{DS3} \approx 0$. As a result, $v_{ss2} \approx v_o^-$. Figure 3.24 sketches the waveform of v_{ss1} and v_{ss2}. It is seen that the frequency of v_{ss1} and v_{ss2} is twice that of v_o^+ and v_o^-. Frequency doubling is therefore achieved.

One important observation is that the voltage swing of v_{ss1} and v_{ss2} is significantly smaller as compared with that of v_o^+ and v_o^-. If v_{ss1} and v_{ss2} are used as one of the mixing signals to realize frequency division of 3, care is needed to ensure that frequency components at $2\omega_o$ are sufficiently large prior to their mixing with the injection signal at $3\omega_o$.

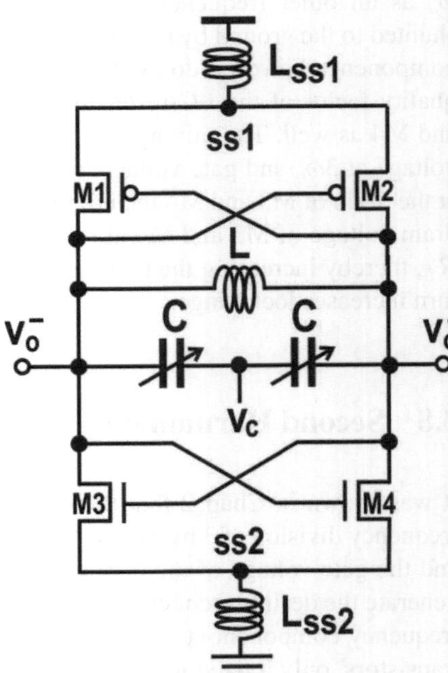

Fig. 3.23 Extraction of the second harmonic of the output of harmonic oscillators—schematic

Fig. 3.24 Extraction of the second harmonic of the output of harmonic oscillators—waveforms

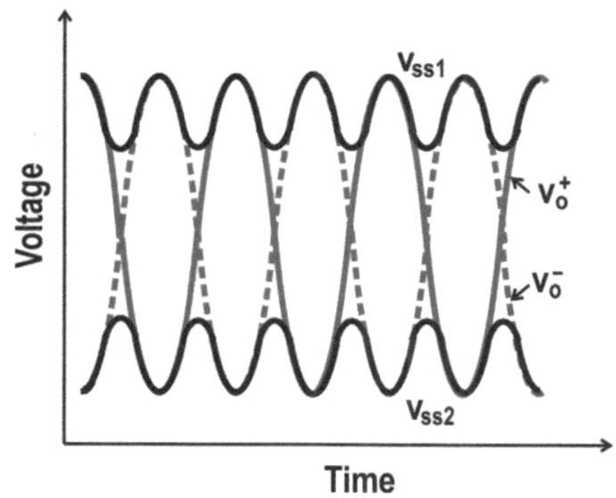

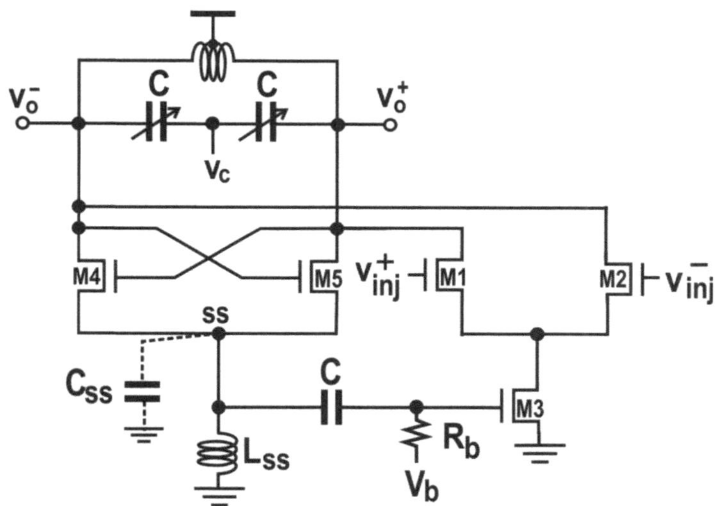

Fig. 3.25 Divide-by-3 ILFD using second harmonic feedback

Figure 3.25 shows a divide-by-3 ILFD that utilizes a second-harmonic feedback technique [78]. The main LC resonator of the ILFD formed by the center-tapped inductor and a pair of varactors resonates at ω_o, which is the frequency of the output of the ILFD. The voltage at the common-source node, denoted by v_{ss}, is at $2\omega_o$ for reasons explained earlier. In order to mix v_{ss} with the injection signal at $3\omega_o$, v_{ss} is applied to the gate of M3. L_{ss} is needed to resonate out the capacitance at the sources of M4 and M5 at $2\omega_o$ so as to maximize the gate voltage of M3. Note that R_b needs to be sufficiently large in order to minimize signal loss. M3 is biased in saturation and functions as a linear transconductor whose output is given by $i_{ds3} = g_{m3}v_{ss2}$. The transconductance of M3 can be adjusted by varying its DC biasing voltage V_b. Transistors M1–M3 form a single-balanced mixer whose output is at ω_o. The frequency components other than ω_o generated in the mixing process are shunted to the ground by the main LC resonator. Note that in single-balanced mixers, we prefer

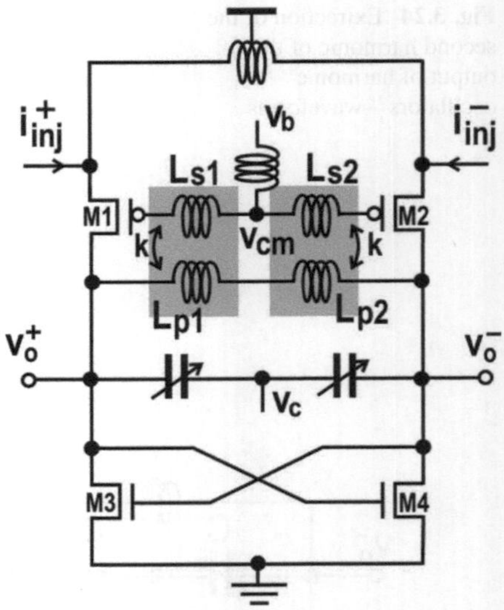

Fig. 3.26 Divide-by-3 ILFD using transformer-based second harmonic feedback

the gating signals v_{inj}^+ and v_{inj}^- to be full-swing square waves such that transistors M1 and M2 operate in an ON/OFF mode. As a result, i_{ds3} is completely steered to M1 or M2 thereby minimizing the loss of i_{ds3}. If v_{inj}^+ and v_{inj}^- are small and M1 and M2 operate in saturation, desired mixing operation will still take place but with a reduced conversion gain subsequently a smaller lock range.

It becomes clear that the operation principle of the divide-by-3 ILDS shown in Fig. 3.25 differs fundamentally from the divide-by-3 ILFD shown in Fig. 2.27. Since frequency mixing is performed by a single-balanced mixer in the divide-by-3 ILDS in Fig. 3.25 rather than utilizing the nonlinear characteristics of transistors in Fig. 2.27, the conversion gain of the mixer, which directly affects the strength of the frequency component at $2\omega_o$ subsequently lock range, can be varied by adjusting V_b.

Another example of utilizing a second harmonic generator in divide-by-3 ILFDs is shown in Fig. 3.26 [32]. Two transformers are used to couple the output voltages v_o^+ and v_o^- to the gates of M1 and M2. The second harmonic is generated as follows: When v_o^+ is sufficiently high while v_o^- is sufficiently low, the gate voltage of M1 will also be high while that of M2 will be low. As a result, M1 will switch off while v_{cm} will follow the drain voltage of M2. Similarly, when v_o^- is sufficiently high while v_o^+ is sufficiently low, the gate voltage of M2 will also be high while that of M1 will be low. As a result, M2 will switch off while v_{cm} will follow the drain voltage of M1. v_{cm} thus peaks twice in each cycle of v_o. The frequency of v_{cm} is therefore twice that of v_o.

The third example of utilizing a second harmonic generator in divide-by-3 ILFDs is shown in Fig. 3.27 [97]. The second harmonic is generated as follows: When v_o^+ is sufficiently high while v_o^- is sufficiently low, M4 will enter triode mode, forcing $v_{cm} \approx v_o^-$. Similarly, when v_o^- is sufficiently high while v_o^+ is sufficiently low, M3

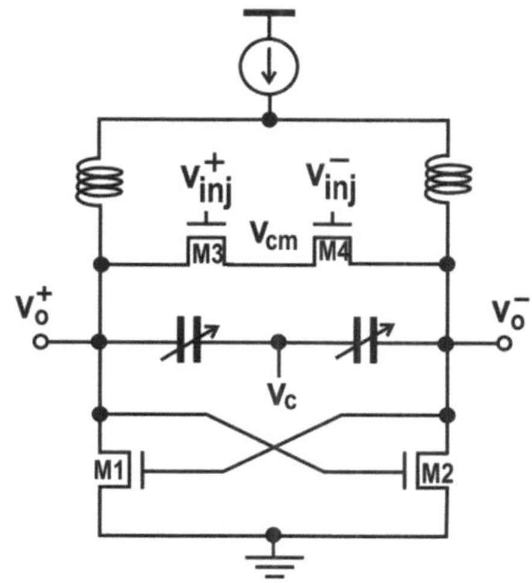

Fig. 3.27 Divide-by-3 ILFD using transistor-based second harmonic feedback

will enter triode mode, forcing $v_{cm} \approx v_o^+$. As a result, v_{cm} ditch twice in one cycle of v_o. The frequency of v_{cm} is therefore twice that of v_o.

When v_o^+ is high while v_o^- is low, M3 will perform frequency mixing with its gate voltage at $3\omega_o$ and source voltage at $2\omega_o$. Similarly, when v_o^- is high while v_o^+ is low, M4 will perform frequency mixing with its gate voltage at $3\omega_o$ and source voltage at $2\omega_o$. The desired frequency component at ω_o resonates at the LC tank while all other frequency components are suppressed by the LC tank.

3.9 Transformer Feedback

It was shown in Sect. 3.3 that in order to increase the lock range of ILFDs using a drain-pumped configuration, the voltage swing of v_{DS} of the injection transistor that performs frequency mixing needs to be maximized. In Fig. 3.11, series-peaking LC tanks are employed at the source and drain of the injection transistor to boost the drain-source voltage v_{DS} of the transistor so as to maximize the lock range [14]. In [18], a divide-by-2 ILFD with transformer feedback was proposed to boost v_{DS} of the injection transistor so as to boost lock range, as shown in Fig. 3.28. The primary windings of transformers 1 and 2 whose inductance is given by L_{1p} and L_{2p}, respectively, form a LC resonator with varactors that resonates at ω_o. Cross-coupled transistors M1 and M2 form a negative resistor to compensate for the resistive loss of the spirals of the primary winding. Note that since $C_{gs1,2}$ and $C_{gd1,2}$, the gate-source and gate-drain capacitors of M1 and M2, respectively, are in parallel with the varactors and are therefore absorbed into the shunt capacitance, M1 and M2 can therefore be considered as ideal transconductors without any capacitance. The coupling between the primary and secondary windings of the transformers forms

Fig. 3.28 Divide-by-2 ILFD with transformer feedback

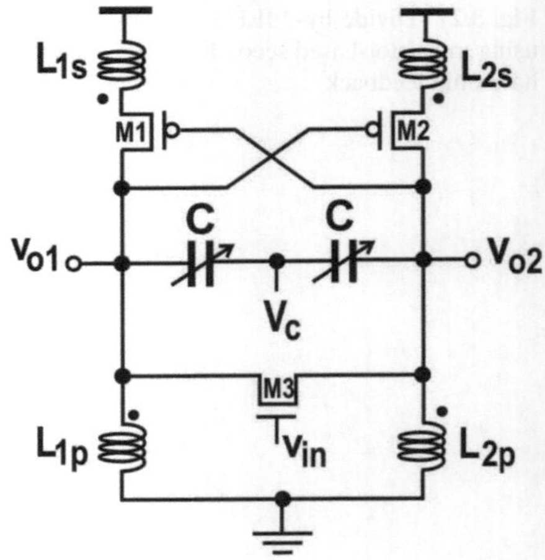

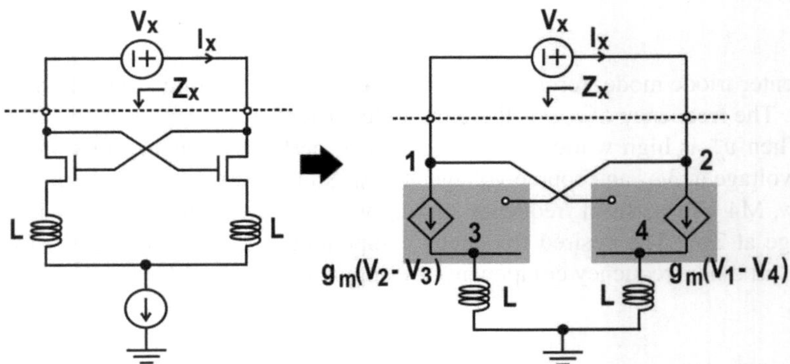

Fig. 3.29 Synthesized negative inductor/resistor. Left: Simplified schematic. Right: Small-signal equivalent circuit with r_o of transistors neglected

positive feedback that enables the voltage of the source and that of the drain of M3 to exceed the supply voltage or drop below the ground potential, thereby maximizing v_{DS3} of the drain-pumped transistor M3. This in turn maximizes the harmonics generated by the mixing operation performed by M3.

As compared with the negative resistor studied in Chap. 2, the negative resistor shown in Fig. 3.28 has inductor source degeneration. To investigate whether this configuration provides a negative impedance or not, let us consider the circuit shown in Fig. 3.29. The small-signal equivalent circuit is also shown in the figure with V_x as a test voltage source. Note that we have assumed that the two transistors are identical and have the same transconductance g_m. We have also neglected the output resistance of the transistors for simplicity. Write Kirchhoff's Current Law at nodes 1–4

$$-I_x = g_m(V_2 - V_3), \tag{3.69a}$$

$$I_x = g_m(V_1 - V_4), \tag{3.69b}$$

$$g_m(V_2 - V_3) = \frac{V_3}{sL}, \tag{3.69c}$$

$$g_m(V_1 - V_4) = \frac{V_4}{sL}, \tag{3.69d}$$

We obtain the input impedance seen into nodes 1 and 2 by solving (3.69)

$$Z_x = \frac{V_x}{I_x} = -2\left(sL + \frac{1}{g_m}\right). \tag{3.70}$$

It is seen from (3.70) that the impedance looking into nodes 1 and 2 is indeed negative. When the inductors are absent, i.e., $L = 0$, Eq. (3.70) will be simplified to

$$Z_x = -\frac{2}{g_m}. \tag{3.71}$$

The circuit in this case synthesizes a negative resistor. When the inductors are presented, since $1/g_m$ is typically small, when $|\omega L| \gg 1/g_m$, the circuit will synthesize a negative inductor.

It is interesting to note that if we replace the source degeneration inductors with capacitors, as shown in Fig. 3.30, the circuit synthesizes a negative capacitor in series with a negative resistor. Note that the circuit is modified in order to provide proper DC biasing currents for the transistors. Note that the node between the two capacitors is an AC ground. Following the same procedures as those for the preceding synthesized negative inductor, one can show that

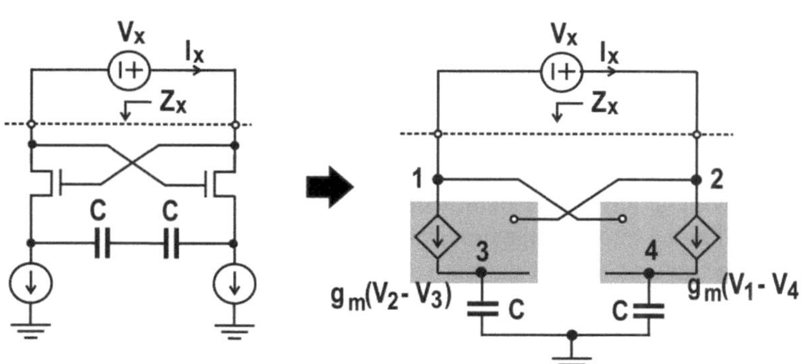

Fig. 3.30 Synthesized negative capacitor/resistor. Left: Simplified schematic. Right: Small-signal equivalent circuit with r_o of transistors neglected

$$Z_x = -2\left(\frac{1}{sC} + \frac{1}{g_m}\right). \tag{3.72}$$

When $|1/\omega C| \gg 1/g_m$, the circuit will synthesize a negative capacitor. Otherwise, it synthesizes a negative capacitor in series with a negative resistor.

A notable characteristic of this ILFD is that in the DC steady-state, only one transistor exists between the power and ground rails, making it particularly attractive for low-voltage applications. As demonstrated in [18], the frequency divider implemented in a 0.18 μm CMOS can operate at 0.75 V even though the nominal supply voltage of a 0.18 μm CMOS technology is 1.8 V typically. The ILFD performs divide-by-2 frequency division with an input at 4 GHz and achieves a lock range of 37%.

3.10 Dual Injections

It was shown in Chap. 2 that divide-by-2 ILFDs can be realized using either tail injection where an injection signal is injected into the sources of the cross-coupled transistors via the tail current source transistor or direct injection where the injection signal is injected directly into the resonators of the ILFDs via the drains of the cross-coupled switching transistors [94]. The former has the advantage that the injection signal does not need to be larger but suffers from the drawback of a low injection efficiency due to the existence of a low-impedance path from the common source node of the cross-coupled switching transistors to the ground, while the latter enjoys the advantage of a high injection efficiency as the injection signal is injected directly into the output LC resonator. In order to increase lock range, Luo and Chen showed that both tail injection and direct injection can be deployed simultaneously, as shown in Fig. 3.31 [45]. The injection signal is applied to both the sources of the cross-coupled switching transistors M3 and M4 and the gate of direct injection transistor M5. The LC tank formed by L_2 and C_2 and resonates at the frequency of the injection signal is needed at the sources of M3 and M4. The LC tank serves two distinct purposes: (a) It provides a current path from the sources of M3–M4 to ground at frequencies other than $\omega_{inj} = 2\omega_o$ where ω_o is the output frequency of the ILFD so that the resonator of the ILFD can oscillate at ω_o. (b) It functions as an open-circuit at $2\omega_o$ such that the injection signal can be applied to the sources of M3–M4 without any loss. Isolation capacitor C_1 prevents the DC current of the ILFD from flowing into the injection signal source. The injection signal can also be applied to the gate of the tail current source transistor M6, as shown in Fig. 3.32. In this case, voltage-to-current conversion is performed by M6. As pointed out earlier in Chap. 2 that inductor L_2 is needed to resonate out C_{ss}, the capacitance encountered at the common source node of the cross-coupled transistors so as to improve injection efficiency. Capacitor C_2 is needed to eliminate the DC path from node ss to the ground. The resonant frequency of the resonator formed by C_2 and L_2 should be much larger than $2\omega_o$.

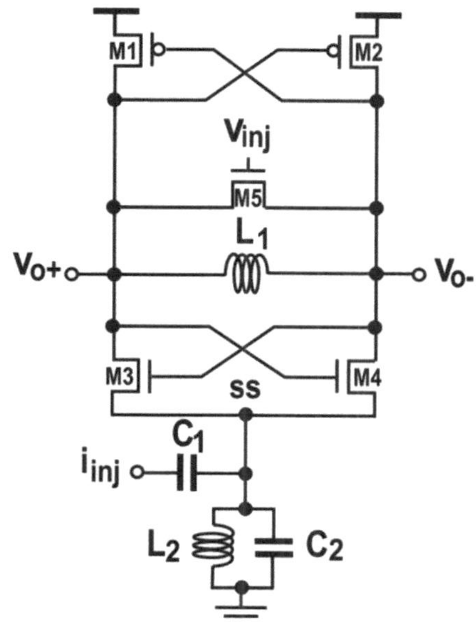

Fig. 3.31 Divide-by-2 ILFD with both tail and direction injections

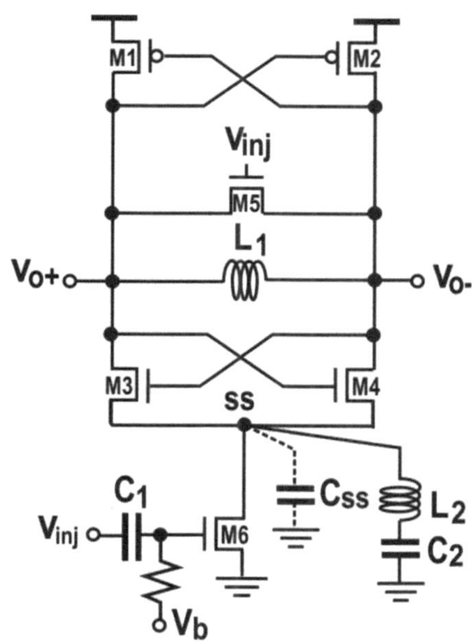

Fig. 3.32 Divide-by-2 ILFD with both tail and direction injections

3.11 Chapter Summary

This chapter studied the design techniques that increase the lock range of harmonic oscillators. We showed that although increasing the strength of injection signals improves the lock range, such an approach might be difficult to implement in some applications such as the injection-lock based frequency calibration of the system clock of a PWM where injection signal sent by the base station of the PWM is typically weak when it reaches the PWM. A more desired approach to increase the lock range of a harmonic oscillator is to modify either the topology or characteristics of the oscillator in the design stage such that the effective injection signal, which is the signal applied to the core of the oscillator at which the desired frequency component is generated, is maximized.

We first investigated the dependence of the lock range of harmonic oscillators on injection signaling. Specifically, we examined the lock range of harmonic oscillators with two unitone injections and the relation between the lock range of the oscillators and the phase of the injection signals when the locations of injections are known. We showed that dual unitone injections yield a larger lock range as compared with single unitone injection provided that the phases of the injection signals are properly chosen. Using a harmonic oscillator with dual unitone injections, we demonstrated both theoretically and numerically that the lock range of the oscillator will be maximized if the injection signals are out of phase and minimized if the injection signals are in phase.

Our focus was then turned to the investigation of the techniques that boost the effective injection signal of harmonic oscillators without modifying external injection signals. We first studied an inductive series-peaking technique to increase the lock range of injection-locked harmonic oscillators using a drain-pumped injection architecture. We showed that the deployment of inductors at the source and drain of the injection transistor where frequency mixing takes place boosts the drain-source voltage of the injection transistor subsequently the impact of the drain-source voltage on the modulation of the nonlinear transconductance of the transistor. This in turn increases the effective injection signal subsequently lock range. Inductor series-peaking was also used in divide-by-3 ILFDs to increase lock range.

The inductive shunt-peaking technique that increases the lock range of injection-locked harmonic oscillators with tail injection was also investigated. We showed that injection efficiency is lowered due to the existence of a large shunt capacitance at the common source node of the cross-coupled mixing transistors. The detrimental impact of the large capacitance at the common source node can be effectively mitigated by the addition of a shunt inductor that resonates out the capacitance at the frequency of the injection signal.

For divide-by-3 ILFDs, the effective differential injection signal can be increased by adding an inductor across the sources of the cross-coupled mixing NMOS transistors or the drains of the cross-coupled PMOS transistors. The added inductor resonates out the capacitances encountered at the common source node or common drain node of the mixing transistors at the injection frequency so as to maximize the effective injection signal prior to its mixing with the output of the ILFDs. The resonating inductor can also be replaced with a transformer so as to increase the effective inductance of both windings of the transformer without increasing the number of the turns of the spirals of the windings of the transformer.

We further studied lock range improvement via lowering the quality factor of the resonators of harmonic oscillators by adding a resistor in series with the spiral of the inductors. We showed that the effectiveness of this technique is constrained by the upper bound of the resistance set by Barkhausen magnitude criterion. We also noted that although the reduction of the quality factor of a harmonic oscillator deteriorates the phase noise of the oscillator, when the oscillator is locked to an injection signal whose phase noise is better than that of the oscillator under injection, the phase noise of the injection-locked oscillator will become comparable to that of the injection signal.

Feedback techniques that boost the lock range of harmonic oscillators were also studied. We showed that although the strength of the injection signal of a harmonic oscillator is typically weak, the output of the oscillator under injection is rather large. The lock range of the oscillator can therefore be improved if the output of the oscillator is routed back to the injection node via either a resistor or a transformer. The former routes the outputs of the oscillator directly to the injection nodes without altering the topology of the oscillator, while the latter feedbacks the output of the oscillator to the drain and source of drain-pumped injection transistor so as to boost the drain-source voltage of the mixing transistor and subsequently the lock range. Resistor feedback features a simple configuration while transformer feedback is particularly attractive for low-voltage applications, however, with a heavy silicon area penalty.

To realize divide-by-3 ILFDs, one can utilize the nonlinearities of mixing transistors directly. This approach typically leads to a small lock range and does not provide designers with sufficient design option to improve lock range. The preferred choice is to generate $2\omega_o$ components directly from the output of the oscillator that is typically large and then mix it with the injection signals whose frequency is at $3\omega_o$. Three second harmonic extraction methods were studied. An attractive characteristic of this approach is that designers can control the strength of the generated second harmonic and subsequently the lock range of oscillators.

Finally a dual-injection technique to increase lock range was studied. We showed that tail injection and direct injection for the realization of frequency division of 2 can be deployed simultaneously to boost the lock range of divide-by-2 ILFDs.

Chapter 4
Injection-Locking of Nonharmonic Oscillators

4.1 Introduction

Nonharmonic oscillators such as ring oscillators and relaxation oscillators differ from their harmonic counterparts with the existence of multiple frequency components at the output of these oscillators, arising from the absence of a frequency selection mechanism in these oscillators. Injection-locked nonharmonic oscillators are encountered in a broad range of applications including the pre-scalar of frequency synthesizers for clock generation in digital systems [14, 22, 44, 51, 55, 77, 82, 100] and PWMs as a local system clock generator that generates the system clock for both RF and baseband blocks of the PWMs [104, 106, 113]. In these applications, injection-locking plays a pivotal role in speeding up frequency-locking process, improving frequency accuracy, and minimizing the phase noise of the oscillators. This chapter is concerned with the fundamentals of the injection-locking of nonharmonic oscillators. Techniques that increase the lock range of nonharmonic oscillators will be investigated in the next chapter.

The chapter is organized as follows: Sect. 4.2 explains the need for injection-locked nonharmonic oscillators in applications such as serial data links and PWMs. Section 4.3 investigates the oscillation of ring oscillators, in particular, how Barkhausen criteria can be used to analyze ring oscillators. The modeling of relaxation oscillators is explored in Sect. 4.4 with a special attention to the distinct nonlinear characteristics of relaxation oscillators as compared with their ring oscillator counterparts. Section 4.5 explores the representation of a nonharmonic oscillator with a set of harmonic oscillators whose frequencies are harmonically related so as to allow linear system analysis tools such as Laplace transform valid for harmonic oscillators and the known knowledge of harmonic oscillators to be utilized in the analysis of nonharmonic oscillators. Section 4.6 is concerned with the fundamentals of Volterra series. In Sect. 4.7, two representative nonlinear elements in integrated circuits, specifically the channel current of MOS transistors and the capacitance of pn-junctions, are studied first. The concept of the Volterra

F. Yuan, *Injection-Locking in Mixed-Mode Signal Processing*,
https://doi.org/10.1007/978-3-030-17364-7_4

elements of a nonlinear element and the Volterra circuits of a nonlinear circuit are then introduced and the process of how to obtain them is provided. Section 4.8 explores the modeling of voltage comparators. In Sect. 4.9, the Volterra circuits of an injection-locked dual-comparator relaxation oscillator are derived and the characteristics of the derived Volterra circuits are investigated. Section 4.10 explores how the Volterra circuit approach can be used to analyze the dual-comparator relaxation oscillator under the injection of a differential current and how the high-order Volterra circuits of the oscillator contribute to the effective injection signals of the first-order Volterra circuit of the oscillator. Section 4.11 investigates the lock range of the dual-comparator relaxation oscillator. The chapter is summarized in Sect. 4.12.

4.2 Injection-Locking of Nonharmonic Oscillators

Unlike harmonic oscillators whose output only contains single frequency component, which is the resonant frequency of the resonator of the oscillators despite the existence of multiple frequency components internally, nonharmonic oscillators distinct themselves from their harmonic counterparts with the existence of multiple frequency components at the output of these oscillators, arising from the absence of a frequency selection mechanism, typically a high-Q resonator, in these oscillators. The absence of costly spiral inductor/transformer resonators in relaxation oscillators greatly reduces the silicon area of these oscillators. In addition, relaxation oscillators feature a large frequency tuning range, accredited to their low-quality factor. This is in sharp contrast to harmonic oscillators, which have a small frequency tuning range typically. Nonharmonic oscillators also feature low power consumption, making them particularly attractive for applications where power consumption is of great importance. Nonharmonic oscillators have found a broad spectrum of applications including data links and PWMs.

The key operations of a serial link transmitter include parallel-to-serial data conversion that serializes parallel data prior to their transmission and pre-emphasis that pre-distorts data objectively so as to counteract the impact of the finite bandwidth of the channels over which data are transmitted. These operations are controlled by the system clock of the transmitter typically generated using a frequency synthesizer with a ring oscillator. In order to stabilize the frequency of the oscillator, a reference clock with a stable frequency is generally used and the ring oscillator is calibrated with respect to the reference clock using a frequency-locked loop. As the reference clock is typically provided by a crystal oscillator whose frequency is typically in a low MHz range while the frequency of the system clock is high, often in a multi-GHz range, a frequency divider with a large division ratio is needed. Although current-mode-logic (CML) based frequency dividers offer great flexibility in terms of division ratio, they are power-greedy and cannot operate at high frequencies [88]. A pre-scalar is needed as the first stage of the chain of frequency dividers to perform frequency division at high frequencies such that CML-based frequency dividers can be used afterwards. Injection-lock based frequency division is particular attractive in implementation of the pre-scalar due to its low power consumption and ability to operate at high frequencies. Injection-locked nonharmonic frequency dividers are generally favored over their harmonic counterparts in digital systems, accredited to their low power consumption, a small silicon area, a large lock range, and moderate jitter constraints. Not only the frequency of the injection signal of the injection-locked nonharmonic frequency divider, which is the output of the ring oscillator of the frequency synthesizer that generates the clock, is subject to the effect of PVT uncertainty, the frequency of the oscillator of the injection-locked frequency divider itself is also affected by PVT uncertainty. As a result, maximizing the lock range of the oscillator of the injection-locked frequency divider is critical [14, 22, 34, 44, 55, 77, 82, 95, 100].

Nonharmonic oscillators are also widely used in ultra low-power wireless systems such as PWMs as a local system clock generator. The frequency of the system clock of PWMs, which controls the operation of both the RF and baseband blocks of the PWMs, needs to be precisely controlled as there are stringent constraints on the frequency of the modulator for up-link and that of the demodulator for down-link. For example, EPC (electronic product code) radio-frequency identity protocols class-1 generation-2 UHF RFID (radio-frequency identification) protocols require that the accuracy of the frequency of the backscattered data be upper-bounded by $\pm 4\%$ [20]. The frequency of the system clock of these systems is prone to the impact of PVT uncertainty. It needs to be calibrated against a reference clock of a known frequency prior to their intended operations. Although frequency calibration can be performed using an FLL, FLL-based frequency calibration generally suffers from a high level of power consumption, a low frequency accuracy, and a long calibration time, making it less attractive for PWMs. Frequency calibration using injection-locking with the frequency reference against which the frequency of the local oscillator is calibrated as the injection signal is an effective way to lower power consumption, improve low frequency accuracy, and shorten calibration time [3, 4, 37, 73].

As compared with FLL-based frequency calibration, injection-locking-based frequency calibration offers three intrinsic advantages:

(a) *High frequency accuracy*: A high frequency accuracy is ensured by the intrinsic characteristics of injection-locking [66]. Although a PLL with a phase/frequency detector (PFD) or an FLL is capable of achieving frequency-locking with a zero static frequency error, the variation of the output of the charge pump of the PLL and FLL[1] will become very small when the difference between the frequency of the reference signal and that of the local oscillator is small, greatly prolonging the locking process. As a result, frequency-locking is often prematurely terminated to trade frequency accuracy for a shorter lock

[1] Type-2 PLLs with a PFD are generally preferred over their type-1 counterparts, i.e., charge-pump-less PLLs as they allow the independent tuning of the loop bandwidth and damping factor of the PLLs [65, 103].

time. Unlike FLL-based frequency-locking, injection-based frequency-locking utilizes the internal mechanism of the oscillator under injection to sense the difference between the frequency of the reference and that of the oscillator with a smaller time constant, thereby enabling a rapid frequency-locking. The sensitivity of the internal frequency sensing can also be adjusted to further improve frequency accuracy.

(b) *Low phase noise*: Although the phase noise of ring oscillators is in general poor as compared with that of their harmonic counterparts with spiral inductors or transformers, when a ring oscillator is clocked to an external clean signal, the phase noise of the injection-locked ring oscillator will become comparable to that of the injection signal [12, 34, 44, 50, 51, 64, 111, 112]. This unique characteristic of injection-locked oscillators is important as it allows the use of low-cost oscillators such as ring oscillators or active-inductor LC oscillators to generate a low phase noise system clock, despite the inferior phase noise performance of these oscillators in their free-running state.

(c) *Fast locking*: The mechanism of injection-locking differs fundamentally from that of PLLs with a PFD or FLLs. The former utilizes the internal mechanism of the oscillator under injection to achieve frequency sensing and adjustment while the latter relies on an external negative feedback loop to adjust the control voltage subsequently the frequency of the oscillator. The time constant of the former is significantly smaller as compared with that of the latter. As a result, the locking process of injection-based frequency-locking is significantly shorter as compared with that of FLL-based frequency-locking.

The success of injection-lock based frequency calibration critically depends upon the lock range of the oscillator under injection especially when the strength of the injection signal is weak. Although increasing the strength of injection signals improves lock range, this approach might not be feasible for some applications. For example, in the injection-lock based frequency calibration of a PWM, the strength of the injection signal, which is the frequency reference sent by the base station of the PWM, decays rapidly with the distance between the PWM whose clock frequency is to be calibrated and its base station. As a result, the strength of the injection signal will be significantly weakened when it arrives at the PWM. On top of that, since the injection signal sent from the base station to the PWM is also the source of the power from which the PWM harvests its operational power, if ASK is used as the data link protocol between the base station and the PWM, a smaller modulation index of the ASK-modulated carrier sent by the base station is preferred in order to minimize the fluctuation of the power harvested by the PWM. This, however, leads to a small injection signal. Increasing the lock range of an injection-locked nonharmonic oscillator without increasing the strength of its injection signal is evidently critical in these applications.

4.3 Modeling of Ring Oscillators

The oscillation of a ring oscillator is initiated by either device noise such as thermal and flicker noise of the transistors generated internally or disturbances coupled externally such as switching noise present on the power and ground rails of the oscillator. In order for the oscillator to start and sustain oscillation, Barkhausen criteria, both magnitude criterion and phase criterion, must be satisfied. To gain an insight on how Barkhausen criteria guide the oscillation of ring oscillators, consider a three-stage ring oscillator represented by the linear negative feedback system shown in Fig. 4.1. To simplify analysis, the delay stages of the oscillator are modeled using a single-pole amplifier whose transfer function is given by

$$A(s) = \frac{A_o}{\frac{s}{\omega_b} + 1}, \tag{4.1}$$

where A_o is the DC gain and ω_b is the -3 dB frequency of the amplifier.

If the delay stages are made of static CMOS inverters shown in Fig. 4.2, in the transition region of the inverter where both the NMOS and PMOS transistors of the inverter operate in saturation, the delay stage can be depicted using a transconductor whose transconductance is given by

$$G_m = g_{m,n} + g_{m,p}, \tag{4.2}$$

where $g_{m,n}$ and $g_{m,p}$ are the transconductance of the NMOS and PMOS transistors, respectively. The output resistance of the transconductors is given by $R = r_{o,n}||r_{o,p}$ where $r_{o,n}$ and $r_{o,p}$ are the output resistance of the NMOS and PMOS transistors, respectively. The load capacitance C consists of three distinct parts, specifically the output capacitance of the driving stage, the input capacitance of the driven stage, and the capacitance of the interconnect connecting the driving and driven stages. The output voltage of the delay stage is given by

$$V_o = \left(R||\frac{1}{sC}\right) G_m V_{in} = \frac{RG_m}{sRC + 1} V_{in}$$

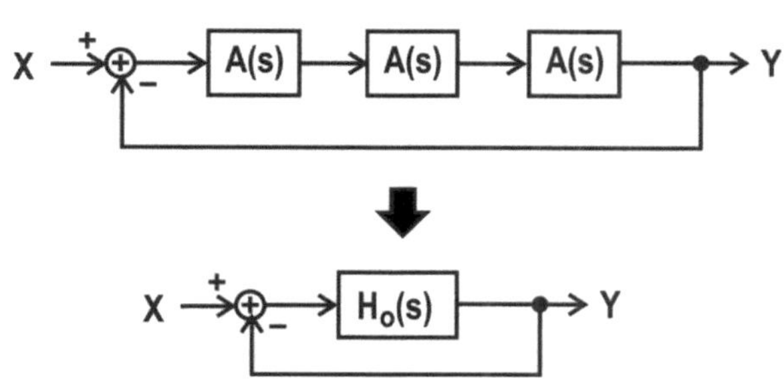

Fig. 4.1 Linear negative feedback system representation of a three-stage ring oscillator in the vicinity of the equilibrium state

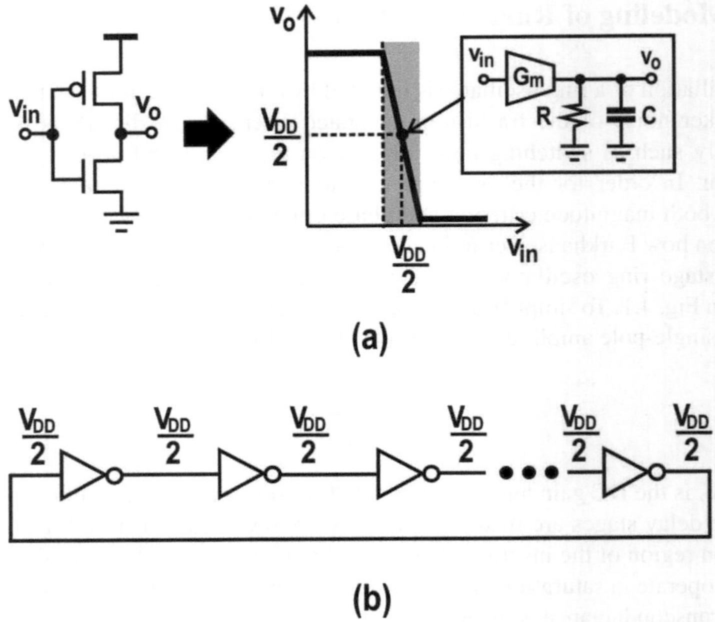

Fig. 4.2 (a) Static CMOS inverter delay stage of ring oscillators. (b) The equilibrium state of ring oscillators

$$= \frac{A_o}{\frac{s}{\omega_b} + 1}. \tag{4.3}$$

We therefore have

$$A_o = RG_m, \tag{4.4}$$

and

$$\omega_b = \frac{1}{RC}. \tag{4.5}$$

Let us now consider a three-stage ring oscillator that has three identical static inverter delay stages. It is seen in Fig. 4.2a that when the input of the inverter is $V_{DD}/2$, the output of the inverter will also be $V_{DD}/2$. This observation reveals that there is an equilibrium state in static inverter ring oscillators occurring when the input and output voltages of the delay stages of the oscillators are at $V_{DD}/2$ ideally, as shown in Fig. 4.2b. If no disturbance is injected into the oscillator in the equilibrium state, the equilibrium state of the oscillator will be retained indefinitely.

When a disturbance is injected into the input of one of the delay stages of the oscillator in the equilibrium state, the large gain of the delay stage in the equilibrium state will force the output of the delay stage to move away from its equilibrium state. The output of the delay stage will be much larger than the injected disturbance due to the gain of the stage in the transition region. It will be further amplified by the following stage and this process continues. The loop gain of the oscillator, however, will not go unbounded as the internal mechanism of the oscillator modulates the transconductance of the delay stages of the oscillator such that the loop gain is adjusted automatically, ensuring Barkhausen criteria is satisfied and oscillation is sustained.

To investigate the start of the oscillation of the oscillator, let us consider the oscillator in its equilibrium state. The gain of the forward path of the ring oscillator in this case is given by

$$H_o(s) = \frac{A_o^3}{\left(\frac{s}{\omega_b}+1\right)^3}. \tag{4.6}$$

In order to satisfy Barkhausen phase criterion, $\angle H_o(j\omega_o) = -180^\circ$ is required. Note that the remaining -180° phase shift required by Barkhausen phase criterion is provided by the negative feedback. To satisfy $\angle H_o(j\omega_o) = -180^\circ$, since all stages are identical, we impose

$$\angle A(j\omega_o) = -60^\circ. \tag{4.7}$$

Utilizing

$$\angle A(j\omega_o) = -\tan^{-1}\left(\frac{\omega_o}{\omega_b}\right), \tag{4.8}$$

we arrive at

$$\omega_o = \sqrt{3}\omega_b. \tag{4.9}$$

Equation (4.9) gives the oscillation frequency of the oscillator. To satisfy Barkhausen magnitude criterion at $\omega_o = \sqrt{3}\omega_b$, we impose

$$\frac{A_o^3}{\left[\sqrt{\left(\frac{\omega_o}{\omega_b}\right)^2+1}\right]^3} = 1. \tag{4.10}$$

Utilizing (4.9), we arrive at

$$A_o = 2. \tag{4.11}$$

Once the constraints specified by (4.9) and (4.11) are satisfied, a small disturbance at the output of stage 1 will be amplified by stage 2. The output of stage 2 will further be amplified by the following stage. This process continues. The existence of positive feedback guaranteed by Barkhausen criteria ensures that the amplitude of the output voltage of the delay stages grows exponentially. When the output voltage of a stage becomes sufficiently large, the PMOS transistor of the driven stage will enter the cut-off mode while the NMOS transistor of the following stage will remain in the saturation mode. The transconductance in this case will drop from $G_m = g_{m,n} + g_{m,p}$ when both transistors are in saturation to $G_m = g_{m,n}$ when the PMOS transistor is in the cut-off while the NMOS transistor is in saturation. Further increasing the output voltage will force the NMOS transistor of the driven stage to enter the triode region and then its transconductance will further drop. It is evident that the input-modulated transconductance of the static inverter delay stages of the oscillator ensures that Barkhausen magnitude criterion will be satisfied when oscillation starts such that oscillation will be sustained.

Figure 4.3 shows the block diagram of the three-stage static inverter ring oscillator. Each static inverter is represented by a nonlinear transconductor consisting of a charging transconductor whose transconductance is given by $G_{m,p}$ and a discharging transconductor whose transconductance is given by $G_{m,n}$, a shunt resistor R representing the output resistance of the transconductors, and a shunt capacitor consisting of the output capacitance of the driving stage, the input capacitance of the driven stage, and the capacitance of the interconnect connecting the driving and driven stages. The transconductor is nonlinear due to the dependence of the transconductance on the input, specifically:

(a) When the input voltage is sufficiently low, the PMOS transconductor will operate in deep triode. $G_{m,p}$ will be low while $G_{m,n} = 0$. The output resistance will be given by $R = r_{on,p}$ where $r_{on,p}$ is the channel resistance of the PMOS transistor. When the input rises, $G_{m,p}$ will rise with the input voltage while $G_{m,n} = 0$ will hold.

(b) When the input voltage exceeds the threshold voltage of the NMOS transistor V_T, $G_{m,n} = g_{m,n}$ will follow. The PMOS transistor in this case will remain in triode and $G_{m,p}$ will be low but higher than that when the PMOS transistor is in

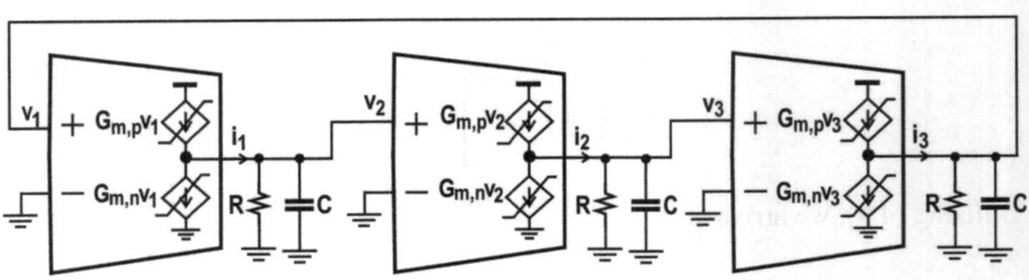

Fig. 4.3 Block diagram of a three-stage ring oscillator. Delay stages are represented by nonlinear transconductors whose transconductance is modulated by the input

deep triode. The output resistance in this case will be given by $R = r_{on,p}||r_{o,n}$ where $r_{o,n}$ is the output resistance of the NMOS transistor.

(c) Further increasing the input voltage will force both transistors to operate in saturation. We therefore have $G_{m,n} = g_{m,n}$ and $G_{m,p} = g_{m,p}$. The output resistance will become $R = r_{o,n}||r_{o,p}$.

(d) When the input voltage is further increased, the PMOS transconductor will continue to operate in saturation while the NMOS transconductor will enter triode. We will therefore have $G_{m,p} = g_{m,p}$ and $G_{m,n}$ will drop below $g_{m,n}$. The output resistance in this case will be given by $R = r_{o,p}||r_{on,n}$ where $r_{on,n}$ is the channel resistance of the NMOS transistor.

(e) Finally when the input voltage exceeds $V_{DD} - |V_{Tp}|$ where V_{Tp} is the threshold voltage of the PMOS transistor, the PMOS transconductor will enter the off state while the NMOS transistor will move to deep triode with its transconductance significantly smaller than $g_{m,n}$. The output resistance in this case will become $R = r_{on,n}$.

Figure 4.4 sketches the transconductance of the NMOS and PMOS transistors of the inverter. The dependence of the transconductance of the transconductors on the input of the inverter is evident.

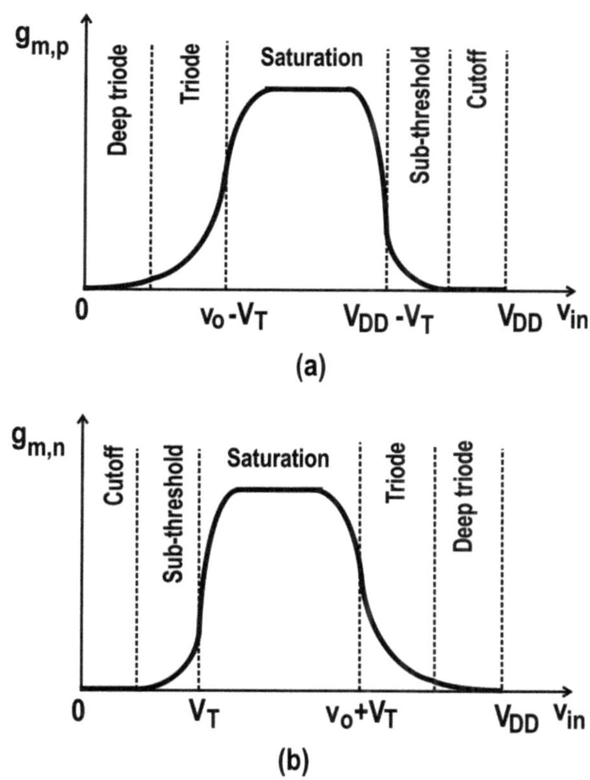

Fig. 4.4 Dependence of the transconductance of NMOS and PMOS transistors of a static inverter on the input voltage of the inverter. (**a**) Transconductance of NMOS transistor. (**b**) Transconductance of PMOS transistor. Assume PMOS and NMOS have the same threshold voltage V_T

4.4 Modeling of Relaxation Oscillators

Unlike inverter-based ring oscillators where the charging and discharging processes of the load capacitor of each delay stage are modulated by the NMOS and PMOS transistors of the driving stage, relaxation oscillators employ comparators to control the charging and discharging processes of the integrating capacitors. The comparators perform both amplitude clamping and phase shifting required to both initiate and sustain oscillation. They exhibit a harsher nonlinear voltage transfer characteristic as compared with static inverters in ring oscillators investigated earlier.

Consider the simplified schematic of a dual-comparator relaxation oscillator shown in Fig. 4.5 [13, 38, 52, 89]. The oscillator consists of a pair of voltage comparators, a set-reset (SR) latch, a pair of identical integrating capacitors, and four digitally modulated current sources.[2] The voltage transfer characteristics of the comparators are shown in Fig. 4.6.

When the oscillator is powered up, the outputs of the SR latch will assume complementary logic states much faster as compared with the voltage of all other nodes of the oscillator due to the regenerative mechanism present in the SR latch. If the SR latch is implemented using a pair of cross-coupled NOR2 gates, the state where S = 1 and R = 1 will be prohibited. This can be ensured by adding a

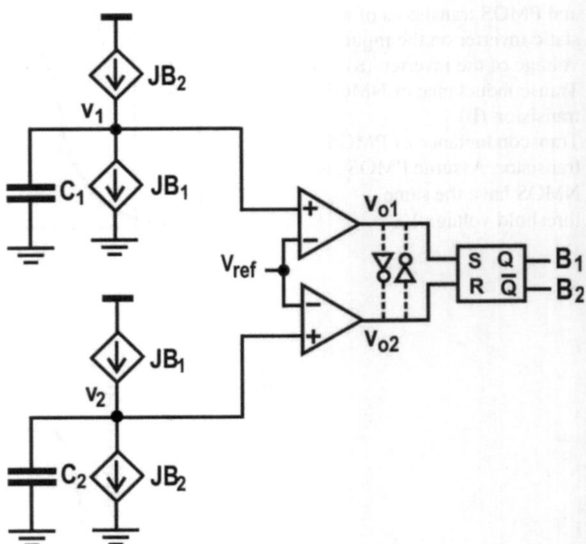

Fig. 4.5 Simplified schematic of dual-comparator relaxation oscillator

[2]Practical relaxation oscillators typically employs four transistors operated in an ON/OFF mode, as shown in Fig. 4.7, rather than current sources to charge and discharge the capacitors so as to greatly simplify design as the voltage of the integrating capacitors does not need to vary with time in a linear fashion.

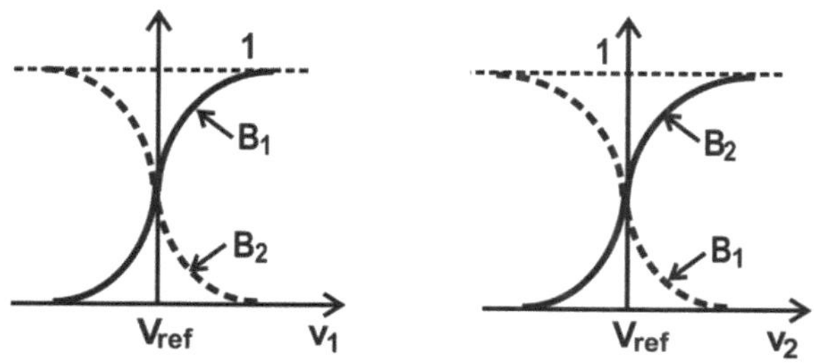

Fig. 4.6 Voltage transfer characteristics of comparators

couple of cross-coupled weak inverters at the input of the SR latch. The addition of the inverter pairs also eliminates the hold state of the SR latch where S $= 0$ and R $= 0$, which is not used in the operation of the oscillator. Similarly, if the SR latch is implemented using a pair of cross-coupled NOR2 gates, the state where S $= 1$ and R $= 1$ will be prohibited. The added inverters also ensure the absence of this prohibited state.

The complementary outputs of the SR latch manage the charging and discharging processes of capacitors C_1 and C_2 by the modulated current sources of constant current J. If $B_1 = 0$ and $B_2 = 1$, v_1 will rise from 0 V while v_2 will remain at 0 V. $v_{o1} = 0$ and $v_{o2} = 0$ will remain unchanged as $v_1 < V_{ref}$ and $v_2 < V_{ref}$. When v_1 exceeds V_{ref}, $B_1 = 1$ and $B_2 = 0$ will be set. As a result, v_2 will rise from 0 V while v_1 will drop from V_{ref}. When v_2 exceeds V_{ref} and v_1 drops below V_{ref}, $B_1 = 0$ and $B_2 = 1$ will be set. This process repeats and oscillation is sustained.

The preceding brief description of the operation of the oscillator reveals the two distinct functionalities of the comparators: (a) Each comparator will provide $-180°$ phase shift when its input surpasses the reference voltage V_{ref}. With two comparators in the loop, a total of $-360°$ phase shift is obtained, thereby satisfying Barkhausen phase criterion. (b) The comparators function as voltage limiters that impose the constraint $v_1 \leq V_{ref}$ and $v_2 \leq V_{ref}$, hence ensuring that Barkhausen magnitude criterion is satisfied.

As an example, consider the dual-comparator relaxation oscillator shown in Fig. 4.7. The comparators are realized using a pair of differential-input single-ended output pre-amplifiers followed by a SR latch. The pre-amplification stages are needed as they enable the comparators to pick up a small difference between v_1 and V_{ref}, and that between v_2 and V_{ref}. The simulated waveform of the voltages of the oscillator is plotted in Fig. 4.8. It is seen that the outputs of the SR latch assume complementary logic states much faster as compared with the voltage of the capacitors due to the regenerative mechanism of the SR latch, ensuring the proper operation, specifically the charging and discharging of the capacitors, of the oscillator. Also noted v_1 and v_2 rise with time linearly. The maximum value of v_1 and v_2 is the reference voltage V_{ref}. The outputs of the oscillator are the output of the SR latch, i.e., Q and $\overline{Q}$, both are square waves with a 50% duty cycle.

Similar to the three-stage static inverter ring oscillator studied earlier, the dual-comparator relaxation oscillator can be represented by the block diagram shown in Fig. 4.9. The comparator and current sources are jointly depicted using two nonlin-

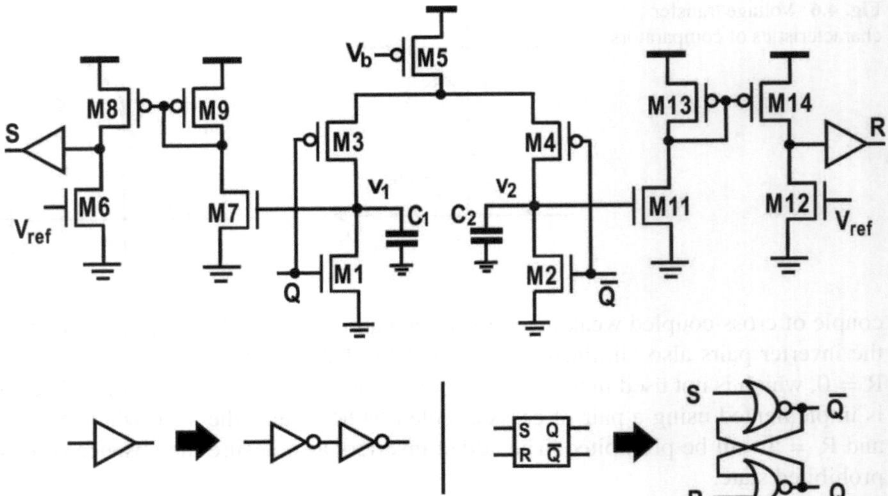

Fig. 4.7 Dual-comparator relaxation oscillator. Technology: TSMC 65 nm 1.2 V CMOS. V_{ref}, $V_b = 0.5$ V

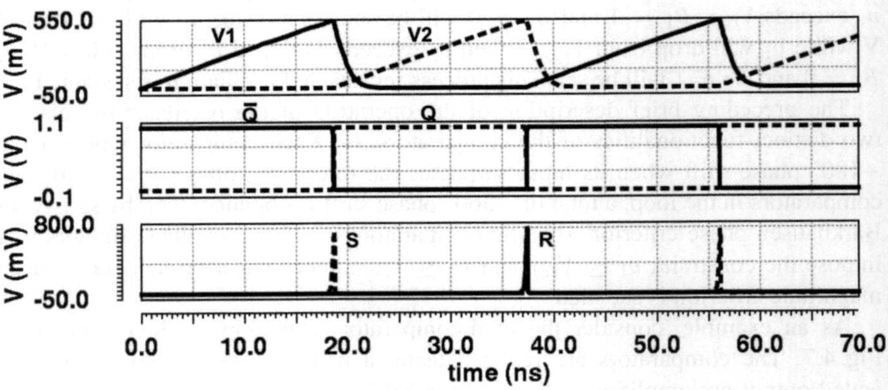

Fig. 4.8 Simulated voltages of dual-comparator relaxation oscillator of Fig. 4.7

ear transconductors whose current is a nonlinear function of $v_n - V_{ref}$, $n = 1, 2$. Similar to the ring oscillator in Fig. 4.3, the transconductance of the transconductors of the relaxation oscillator is modulated by the input of the transconductors. The degree of the nonlinear characteristics of the transconductors of the relaxation oscillator, however, is higher as compared with that of the ring oscillator, rooted to the regenerative mechanism of the SR latch that is part of the comparators. Unlike the transconductance of the transconductors of the ring oscillator that varies gradually with the input of the transconductors, the transconductance of the

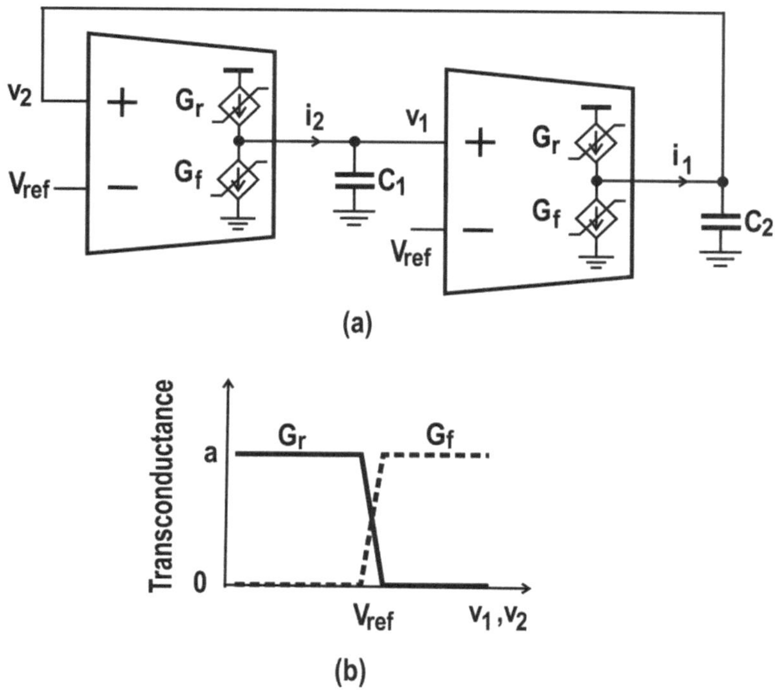

Fig. 4.9 (**a**) Block diagram of dual-comparator relaxation oscillator. (**b**) Transconductance of the transconductors

transconductors of the relaxation oscillator remains unchanged until the input of the transconductors crosses threshold voltage V_{ref}. Specifically, when $v_1 < V_{ref}$, the transconductance of the charging transconductor G_r will be constant, i.e., $G_f = a$ where a is a constant while that of the discharging transconductor G_f will be 0. When $v_1 > V_{ref}$, we will have G_r=0 and $G_f = a$, as shown graphically in Fig. 4.9. The same holds for the other transconductor. Clearly a harsh nonlinear characteristic exists at the threshold-crossing where $v_1, v_2 = V_{ref}$. Also, the transconductances are constrained by the following complementary relation

$$G_r + G_f = a. \tag{4.12}$$

The preceding analysis shows that although both ring oscillators and relaxation oscillators are nonharmonic oscillators, they have a rather distinct characteristic, specifically, the different degrees of the nonlinear characteristics of the transconductors. It is indeed this distinct characteristic that gives rise to the different lock ranges of these oscillators, as to be detailed in the next chapter.

4.5 Representation of Nonharmonic Oscillators

The existence of multiple frequency components in nonharmonic oscillators greatly increases the level of difficulty in analysis of these oscillators. This is because analysis tools valid for analyzing linear systems such as single-variable Laplace transform cannot be applied directly to nonharmonic oscillators. We showed in the previous chapters that a harmonic oscillator can be analyzed as a linear system due to the existence of only a single frequency variable at the input and output of the oscillator, specifically the center frequency of the high-Q resonator of the oscillator while all other frequency components in the oscillator are suppressed by the resonator before they reach the output of the oscillator. It is desirable to explore whether we can utilize familiar linear system analysis tools and the known knowledge of harmonic oscillators to analyze nonharmonic oscillators. This is the focus of this section.

Unlike harmonic oscillators whose output is a sinusoid, the output of a nonharmonic oscillator is ideally a square wave. For example, the waveform of the output of a ring oscillator with static inverter delay stages is approximately a square wave if the charging and discharging times of the load capacitor of each delay stage is sufficiently small as compared with the period of the oscillator. For relaxation oscillators such as the one shown in Fig. 4.7 whose output voltage waveforms are given by Fig. 4.8 (Q and $\overline{Q}$) and are square waves. The output voltage of a nonharmonic oscillator thus contains an infinite number of frequency components as the output of these oscillators can be represented by its Fourier series with an infinite number of frequency components[3] This differs fundamentally from harmonic oscillators whose output is a sinusoidal wave and only contains a single frequency component.

Assume that the waveform of the output of a nonharmonic oscillator is shown in Fig. 4.10. The Fourier series of v_o is given by

$$v_o(t) = \frac{V_m}{2} + \frac{4V_m}{\pi} \sum_{k=1,3,5,\ldots}^{\infty} \frac{1}{k} \sin(k\omega_o t), \tag{4.13}$$

where $\omega_o = 2\pi/T$ and T is the period of the first harmonic of v_o. It is seen that only odd harmonics exist in the spectrum of the output of the oscillator. Utilizing Euler's formula, we can write (4.13) as

$$v_o(t) = \frac{V_m}{2} + \frac{2V_m}{j\pi} \sum_{k=1,3,5,\ldots}^{\infty} \frac{1}{k} \left(e^{jk\omega_o t} - e^{-jk\omega_o t} \right), \tag{4.14}$$

[3]More generally, the waveform of the output of a nonharmonic oscillator takes the shape of a trapezoid. The output can also be represented its Fourier series and contains an infinite number of frequency components.

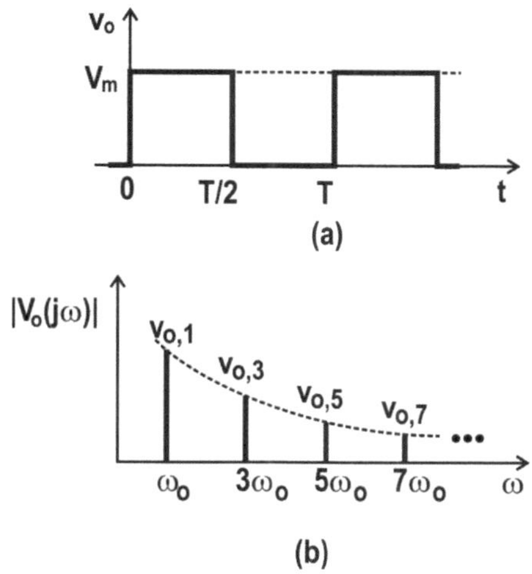

Fig. 4.10 Ideal output of nonharmonic oscillators. (**a**) Time-domain waveform. (**b**) Frequency-domain spectrum

The Fourier transform of the time-varying part of v_o, denoted by $\hat{V}_o(\omega)$, is given by

$$\hat{V}_o(\omega) = \frac{4V_m}{j} \sum_{k=1,3,5,\ldots}^{\infty} \frac{1}{k} \left[\delta(\omega - k\omega_o) - \delta(\omega + k\omega_o)\right], \tag{4.15}$$

where $\delta(x)$ is the unit impulse function defined as

$$\delta(x) = \begin{cases} 1 & \text{if } x = 0 \\ 0 & \text{otherwise.} \end{cases} \tag{4.16}$$

The spectrum of the output of the oscillator consists of a train of impulses at frequencies $\omega_o, 3\omega_o, 5\omega_o, \ldots$, whose amplitude drops with the order of the harmonic in a hyperbolic manner, as shown in Fig. 4.10.[4]

The fact that the output of an ideal nonharmonic oscillator contains tones at ω_o, $3\omega_o$, $5\omega_o$,... suggests that the nonharmonic oscillator can be considered as the assembly of a set of harmonic oscillators that oscillate at ω_o, $3\omega_o$, $5\omega_o$,... Since the output of each of these harmonic oscillators only contains a single frequency component that is harmonically related to others, i.e., no overlapping spectrum, each harmonic oscillator can therefore be considered as a stand-alone linear system and analyzed using linear system approaches. The output of the harmonic oscillators can be added together to yield the output of the nonharmonic oscillator

[4]The frequency components of $\hat{V}_o(\omega)$ at negative frequencies are not shown in Fig. 4.10.

$$\hat{V}_o(\omega) = \sum_{k=1,3,5,...}^{\infty} V_{o,k}(k\omega_o)\delta(\omega - k\omega_o), \tag{4.17}$$

where

$$V_{o,k}(k\omega_o) = \frac{4V_m}{k}. \tag{4.18}$$

Figure 4.11 gives the block diagram representation of the nonharmonic oscillator using a set of harmonic oscillators whose frequency is harmonically related. Each harmonic oscillator is an autonomous system if no external injection signal is present, $v_{in,k} = 0$ for $k = 1, 3, 5 \ldots$.

If an injection signal is injected into the oscillator, each harmonic oscillator will have its own injection signal present at its input, as to be shown shortly. The representation of a nonharmonic oscillator with a set of harmonic oscillators allows the use of analysis methods for harmonic oscillators to analyze nonharmonic oscillators. It also allows us to utilize the known knowledge of harmonic oscillators to better explore the unknowns of nonharmonic oscillators. Although this approach is postulated from a time-domain waveform and frequency-domain spectrum point of view, we will show theoretically in the next section that such a treatment is indeed valid.

When the thermal noise of the devices of the oscillator is accounted for, the waveform of the nonharmonic oscillator will take the shape shown in Fig. 4.12. Note that we are only concerned with the uncertainty of the transition edges of the output of the oscillator as it affects the jitter or phase noise of the oscillator. The uncertainty of threshold-crossing due to the fluctuation of the output voltage of the oscillators arising from device noise gives rise to the fluctuation of the frequency the oscillator. If we neglect the fluctuation of the amplitude of the output voltage when the output of the oscillator is at either high or low as it has a negligible impact on the period of the oscillator, the output voltage of the oscillator can be written as

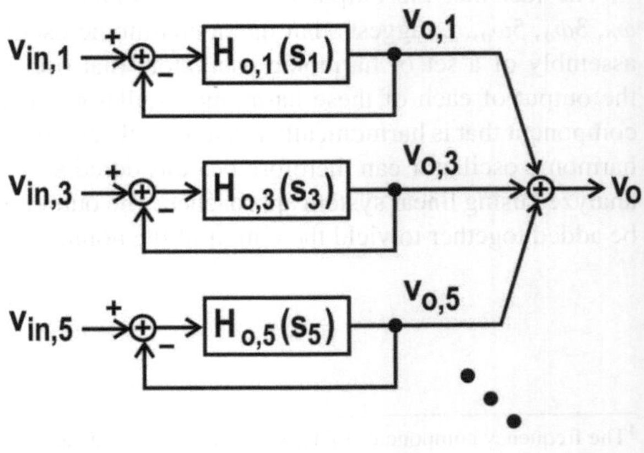

Fig. 4.11 Representation of a nonharmonic oscillator with a set of harmonic oscillators whose frequencies are harmonic related. $s_k = jk\omega$, $k = 1, 3, 5, \ldots$

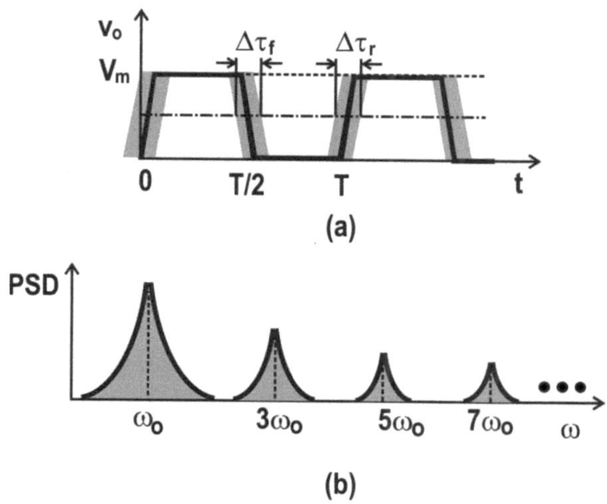

Fig. 4.12 Output of nonharmonic oscillators with noise accounted for. (**a**) Time-domain waveform. (**b**) Frequency-domain spectrum

$$v_o(t) = \frac{V_m}{2} + \frac{4V_m}{\pi} \sum_{k=1,3,5,...}^{\infty} \frac{1}{k} \sin\left[k\omega_o\left(1 - \frac{\Delta T}{T}\right)t\right], \tag{4.19}$$

where ΔT is the variation of the period of the first harmonic of the output of the oscillator caused by device noise. Frequency uncertainty caused by device noise is evident in (4.19) and is given by

$$\Delta\omega = \left(\frac{\Delta T}{T}\right)\omega_o. \tag{4.20}$$

The frequency variation of the kth harmonic of the output of the oscillator is given by

$$k\Delta\omega = k\left(\frac{\Delta T}{T}\right)\omega_o. \tag{4.21}$$

Let the slope of v_o at the threshold-crossing of the rising and falling transitions be S_r and S_f, respectively. Further let the device noise be dominated by the thermal noise and is given by v_n. Utilizing Fig. 4.13, one can show that the difference between the threshold-crossing time of v_n when no device noise is accounted for to that when device noise is accounted for is given by

$$\Delta\tau_r = \frac{v_n}{S_r}, \tag{4.22a}$$

$$\Delta\tau_f = \frac{v_n}{S_f}. \tag{4.22b}$$

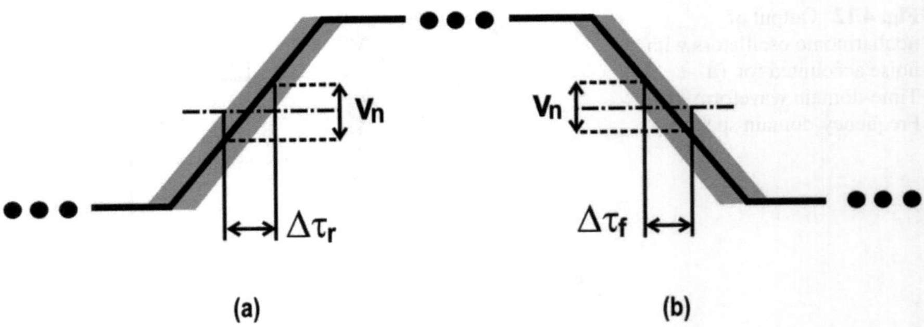

Fig. 4.13 Noise induced timing jitter at threshold-crossings. (**a**) Rising edge; (**b**) falling edge

Since

$$\Delta T = \Delta\tau_r + \Delta\tau_f \tag{4.23}$$

and noting that $\Delta\tau_r$ and $\Delta\tau_f$ are assumed to be uncorrelated as they are rooted to the thermal noise of the devices, we arrive at the variance of the period of the oscillator

$$\begin{aligned}\overline{\Delta T^2} &= \overline{\Delta\tau_r^2} + \overline{\Delta\tau_f^2} \\ &= \left(\frac{1}{S_r^2} + \frac{1}{S_f^2}\right)\overline{v_n^2}.\end{aligned} \tag{4.24}$$

It follows from (4.20) that

$$\overline{\Delta\omega^2} = \left(\frac{\omega_o}{T}\right)^2 \left(\frac{1}{S_r^2} + \frac{1}{S_f^2}\right)\overline{v_n^2}. \tag{4.25}$$

Similarly, for the kth harmonic of the output of the oscillator, the variance of the frequency is given by

$$\overline{k\Delta\omega^2} = \left(\frac{k\omega_o}{T}\right)^2 \left(\frac{1}{S_r^2} + \frac{1}{S_f^2}\right)\overline{v_n^2}. \tag{4.26}$$

Equation (4.25) shows that the power of frequency uncertainty or phase noise of the oscillator is proportional to the power of the device noise and inversely proportional to the square of the slope of transition edges of the output of the oscillator. Since v_n is white, $\Delta\omega$ is white as well. It should be noted that although $\Delta\omega$ is white in nature, the spectrum of the phase noise of the oscillator at ω_o and harmonic frequencies rather follow a skirt profile, rooted to the profile of the frequency response of the resonator that shapes the spectrum of the phase noise of the oscillator.

4.6 Fundamentals of Volterra Series

It was shown in the preceding sections that a distinct characteristic of nonharmonic oscillators is their higher degree of nonlinearity as compared with harmonic oscillators. A nonharmonic oscillator can be represented by a set of harmonic oscillators whose frequencies are the first, third, fifth, ... harmonic of the output of the nonharmonic oscillator.

It is well known that Volterra series, in memory of Italian mathematician and physicist Vito Volterra (1860–1940), is an effective means to analyze nonlinear time-varying systems [67]. A distinct characteristic of Volterra series-based analysis of nonlinear systems that departs it from power series-based analysis of nonlinear systems is its permission of nonlinear elements with memory such as capacitors and inductors whose present response is a function of not only the present input but also the accumulated past response in the systems. The degree of the nonlinearity can be accounted for conveniently by properly truncating the Volterra series expansion of the variables of the nonlinear systems. As the degree of complexity increases exponentially with the increase in the order of the Volterra series expansion of the variables of the nonlinear systems, the order of Volterra series expansion is typically limited to 3. As a result, Volterra series approach is most used in analyzing nonlinear systems with mildly nonlinear characteristics where the third-order Volterra series expansion of the variables of the systems depicts the systems adequately.

The response of a nonlinear system $y(t)$ relates to its input $x(t)$ by a converging Volterra series [86]

$$y(t) = \sum_{n=1}^{\infty} y_n(t), \tag{4.27}$$

where

$$y_n(t) = \int_{-\infty}^{\infty} \cdots \int_{-\infty}^{\infty} h_n(t, \tau_1, \tau_2, \ldots, \tau_n) \prod_{k=1}^{n} x(\tau_k) d\tau_k, \tag{4.28}$$

and $h_n(t, \tau_1, \ldots, \tau_n)$ is the nth-order Volterra kernel. Note that τ is the time at which input $x(t)$ is applied to the system and t is the time instant at which the output of the system is measured. The nonlinear system can be either time-varying or time-invariant. For a time-varying nonlinear system, Eq. (4.27) is used. For a time-invariant system, the output of the system only depends upon the time difference between the time instant at which the input is applied to the system to the time instant at which the output is measured

$$y_n(t) = \int_{-\infty}^{\infty} \cdots \int_{-\infty}^{\infty} h_n(t - \tau_1, t - \tau_2, \ldots, t - \tau_n) \prod_{k=1}^{n} x(\tau_k) d\tau_k. \tag{4.29}$$

If the input of the system is perturbed such that the input is changed from $x(t)$ to $\epsilon x(t)$ where ϵ is a constant and the induced variation of the amplitude of the input is sufficiently small such that the resultant response of the system remains converged, the response will be given by

$$y(t) = \sum_{n=1}^{\infty} y_n(t)\epsilon^n. \tag{4.30}$$

Equation (4.30) reveals that when the amplitude of the input varies, the response of the system will vary accordingly and can be quantified by a power series in ϵ with $y_n(t)$ as the coefficient of the nth-order term of the power series. The purpose of introducing a small perturbation in the input is to allow us to investigate the characteristics of the nonlinear system by examining the response of the system to the perturbation without worrying about the stability of the system.

4.7 Volterra Element and Volterra Circuits

To make use of Volterra series to analyze a nonlinear circuit, we first formulate the governing equations of the circuit that depict the behavior of the circuit. These equations are typically formulated from Kirchhoff's Current Law, also known as nodal analysis[5]

$$f(v_1, v_2, \ldots, v_N, v_{in}) = 0, \tag{4.31}$$

where v_n is the voltage of node n and N is the number of the nodes of the circuit. The nonlinear elements of the circuit are included in the equations and are represented by their Taylor series expansion to the order of 3 typically. For example, the channel current of a NMOS transistor operating at pinch-off is given by

$$\begin{aligned} i_{DS} &= \frac{1}{2}\mu_n C_{ox}\left(v_{gs} + V_{GS} - V_T\right)^2 \\ &= \frac{1}{2}\mu_n C_{ox}(V_{GS} - V_T)^2\left(1 + \frac{v_{gs}}{V_{GS} - V_T}\right)^2 \\ &= \frac{1}{2}\mu_n C_{ox}(V_{GS} - V_T)^2\left[1 + \frac{2v_{gs}}{V_{GS} - V_T} + \frac{v_{gs}^2}{(V_{GS} - V_T)^2}\right], \end{aligned} \tag{4.32}$$

[5] In general, modified nodal analysis that permits branch currents, along with nodal voltages, is used to formulate the governing equations of an arbitrary circuit [85].

where v_{gs} and V_{GS} are the AC and DC components of the gate-source voltage of the transistor, respectively. It can be shown that the AC portion of the channel current, denoted by i_{ds}, is given by

$$i_{ds} = g_{m1} v_{in} + g_{m2} v_{in}^2, \tag{4.33}$$

where

$$g_{m1} = \mu_n C_{ox}(V_{GS} - V_T), \tag{4.34a}$$

$$g_{m2} = \frac{1}{2}\mu_n C_{ox}. \tag{4.34b}$$

For a reverse biased pn-junction with AC and DC biasing voltages v_r and V_R, respectively, its capacitance is given by

$$\begin{aligned} C_J &= \frac{C_{Jo}}{\sqrt{1+\frac{v_r+V_R}{\phi_o}}} \\ &= \frac{C_{Jo}}{\sqrt{1+\frac{V_R}{\phi_o}}} \frac{1}{\sqrt{1+\frac{v_r}{V_R+\phi_o}}} \\ &= \frac{C_{Jo}}{\sqrt{1+\frac{V_R}{\phi_o}}} \left[1 - \frac{1}{2}\left(\frac{v_r}{V_R+\phi_o}\right) + \frac{3}{8}\left(\frac{v_r}{V_R+\phi_o}\right)^2 + \cdots \right], \end{aligned} \tag{4.35}$$

where C_{Jo} is the capacitance with no reverse biasing voltage and ϕ_o is the built-in potential of the pn-junction. The AC portion of the capacitance is given by

$$C_j = a_1 v_r + a_2 v_r^2 + \cdots, \tag{4.36}$$

where

$$a_1 = -\frac{1}{2}\frac{C_{J,DC}}{V_R+\phi_o}, \tag{4.37a}$$

$$a_2 = \frac{3}{8}\frac{C_{J,DC}}{(V_R+\phi_o)^2}, \tag{4.37b}$$

$$C_{J,DC} = \frac{C_{Jo}}{\sqrt{1+\frac{V_R}{\phi_o}}}. \tag{4.37c}$$

Once the governing equations of the circuit are obtained, a perturbation in the input of the circuit x is introduced such that the input of the circuit becomes ϵx. The nodal voltages of the circuit are then represented by their Volterra series expansions to the order of 3. Since the input of the circuit is now ϵx, the voltage of node n is also a third-order polynomial in ϵ with the coefficients to be determined

$$v_n \approx v_{n,1}\epsilon + v_{n,2}\epsilon^2 + v_{n,3}\epsilon^3. \tag{4.38}$$

$v_{n,1}$, $v_{n,2}$, and $v_{n,3}$ are the first-, second-, and third-order terms of the Volterra series expansion of v_n, respectively. Equation (4.38) is substituted into the governing equation (4.31) of the circuit, yielding the following polynomial in ϵ

$$\begin{aligned}&f_1(v_{1,1}, \ldots, v_{N,1}, v_{in})\epsilon\\ &\quad+ f_2(v_{1,1}, \ldots, v_{N,1}, v_{1,2}, \ldots, v_{N,2})\epsilon^2\\ &\quad+ f_3(v_{1,1}, \ldots, v_{N,1}, v_{1,2}, \ldots, v_{N,2}, v_{1,3}, \ldots, v_{N,3})\epsilon^3 \approx 0,\end{aligned} \tag{4.39}$$

where $v_{n,k}$, $k = 1, 2, 3$, is the kth order term of the voltage of node n. Note that we have neglected the terms whose order exceeds 3 in (4.39). Equation (4.39) is a third-order polynomial in ϵ whose value is 0. In order for the value of the polynomial to be 0 for arbitrary ϵ, all the coefficients of the polynomial must also be 0. A set of new governing equations derived by setting the coefficients of the polynomial in (4.39) to 0 are therefore obtained.

$$f(v_{1,1}, \ldots, v_{N,1}, v_{in}) = 0, \tag{4.40a}$$

$$f_2(v_{1,1}, \ldots, v_{N,1}, v_{1,2}, \ldots, v_{N,2}) = 0, \tag{4.40b}$$

$$f_3(v_{1,1}, \ldots, v_{N,1}, v_{1,2}, \ldots, v_{N,2}, v_{1,3}, \ldots, v_{N,3}) = 0, \tag{4.40c}$$

The circuit governed by the first equation of (4.40) obtained by setting the coefficients of the first-order term of the polynomial in (4.39) to 0 is termed first-order Volterra circuit. Similarly, the circuits governed by the second and third equations of (4.40) obtained by setting the coefficients of the second- and third-order terms of the polynomial in (4.39) to 0 are termed second- and third-order Volterra circuits, respectively.

To illustrate how the Volterra circuits of a nonlinear circuit are derived using the preceding approach, consider a nonlinear element characterized by

$$y = a_1 x + a_2 x^2 + a_3 x^3, \tag{4.41}$$

where x and y are the input and output of the element, respectively, and a_1, a_2, and a_3 are constants.[6] Represent x and y in their third-order Volterra series expansion

[6]The Volterra series expansion of a nonlinear circuit accounts for the impact of the elements with memory such as capacitors and inductors by representing the nodal voltages and branch currents in the form of Volterra series expansion of the nodal voltages and branch currents of the circuit with frequency-dependent coefficients. Nonlinear elements are depicted using their Taylor series expansion.

$$x = x_1\epsilon + x_2\epsilon^2 + x_3\epsilon^3, \tag{4.42a}$$

$$y = y_1\epsilon + y_2\epsilon^2 + y_3\epsilon^3, \tag{4.42b}$$

where x_k and y_k, $k = 1, 2, 3$, are the kth-order Volterra series expansion of x and y, respectively. Note that although the nonlinear element depicted by (4.41) has no memory, the impact of elements with memory is accounted for in the Volterra series expansion of circuit variables in (4.42). Substitute (4.42) into (4.41)

$$\begin{aligned} y_1\epsilon + y_2\epsilon^2 + y_3\epsilon^3 &= a_1(x_1\epsilon + x_2\epsilon^2 + x_3\epsilon^3) \\ &+ a_2(x_1\epsilon + x_2\epsilon^2 + x_3\epsilon^3)^2 \\ &+ a_3(x_1\epsilon + x_2\epsilon^2 + x_3\epsilon^3)^3. \end{aligned} \tag{4.43}$$

Equating the terms that have the same order in ϵ yields

$$y_1 = a_1x_1, \tag{4.44a}$$

$$y_2 = a_1x_2 + a_2x_1^2, \tag{4.44b}$$

$$y_3 = a_1x_3 + a_2x_1x_2 + a_3x_1^3. \tag{4.44c}$$

Equation (4.44) characterizes the elements derived using the Volterra series approach and are termed the Volterra elements of the nonlinear element. Since up to the third-order Volterra series expansion is used, for each nonlinear element, there exist first-, second-, and third-order Volterra elements. It is seen from (4.44) that in order to quantify the behavior of the second-order Volterra element of the nonlinear element, the behavior of the first-order Volterra element of the nonlinear element must be already known. Similarly, both the behavior of the first- and second-order Volterra elements of the nonlinear element need to be known in order to quantify the behavior of the third-order Volterra element of the nonlinear element. Note for a linear element, its Volterra element is itself.

Equation (4.44) can be rearranged to the following form

$$y_1 - a_1x_1 = 0, \tag{4.45a}$$

$$y_2 - a_1x_2 = a_2x_1^2, \tag{4.45b}$$

$$y_3 - a_1x_3 = a_2x_1x_2 + a_3x_1^3. \tag{4.45c}$$

It is seen from (4.45) that the expressions on the left-hand side are identical while those on the right-hand side differ. Those on the left-hand side can be considered as the governing equations that depict the linear portion of the Volterra elements of the nonlinear element while those on the right-hand side can be considered as the inputs of the linear portion of the Volterra elements that capture the nonlinear characteristics of the nonlinear element. The Volterra circuits are therefore identical except their inputs. Once the Volterra elements of all linear and nonlinear elements of the nonlinear circuit are available, the nth-order Volterra circuit is constructed by replacing all nonlinear elements with their nth-order Volterra counterpart while leaving all linear elements unchanged. The Volterra circuits of a nonlinear circuit derived this way possess the following important characteristics:

(a) Volterra circuits are linear and have the same topology, element type, and element value except their inputs. They therefore have the same transfer function.
(b) The first-order Volterra element does not have an input associated with it. If the circuit has an input, the actual input of the circuit will only appear in the first-order Volterra circuit.
(c) The input of the second-order Volterra circuit arising from the nonlinearity of the oscillator is a function of the response of the first-order Volterra circuit. This function is nonlinear as it captures the second-order nonlinear characteristics of the circuit.
(d) The input of the third-order Volterra circuit also arising from the nonlinearity of the oscillator is a function of the response of the first- and second-order Volterra circuits. This function is also nonlinear as it captures the third-order nonlinearity of the circuit.
(e) Volterra circuits are solved in a sequential manner with the first-order Volterra circuit solved first. Once the response of the first-order Volterra circuit is available, the input of the second-order Volterra circuit can be obtained and the response of the second-order Volterra circuit can be solved subsequently. This process continues to the third-order Volterra circuit.

4.8 Modeling of Comparators

Voltage comparators feature a harsh nonlinear voltage transfer characteristic as compared with the static inverter stages of ring oscillators. The harsh nonlinear characteristic is rooted to the regenerative mechanism of the SR latch of the comparators. The simplified schematic of the dual-comparator relaxation oscillator with dual current injections is shown in Fig. 4.14. The comparators are realized using a pair of differential amplifiers and a SR latch. The differential amplifiers are needed to enable the comparators to have the ability to sense a small difference between the input voltage and reference voltage V_{ref}. Since the gain of the differential amplifiers is rather small especially at high frequencies, a regenerative SR latch is needed to generate a pair of full-swing complementary Boolean outputs B_1 and B_2, which are used to modulate the charging and discharging current sources of the integrating capacitors. As mentioned earlier a pair of weak static inverters could be added at the input of the SR latch to eliminate input states R, S $=$ 0 and R, S $=$ 1 of the SR latch so as to ensure the proper operation of the oscillator.

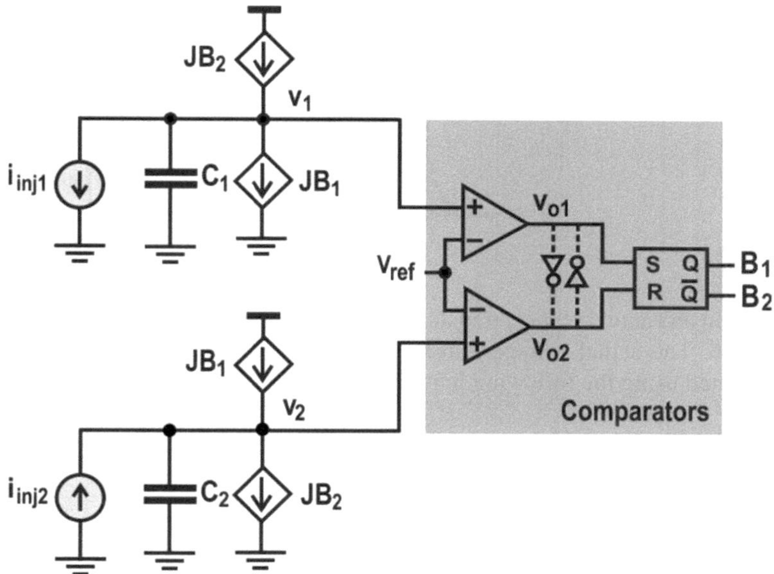

Fig. 4.14 Simplified schematic of dual-comparator relaxation oscillator. i_{inj1} and i_{inj2} are two external currents injected into the oscillator

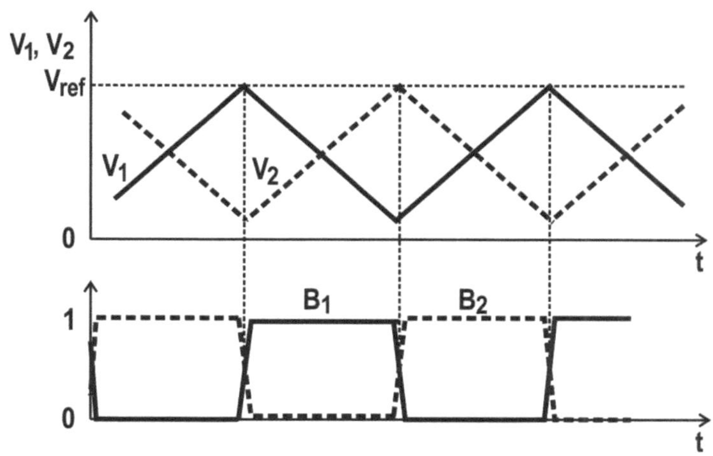

Fig. 4.15 Waveforms of the voltages of dual-comparator relaxation oscillator

Figure 4.15 sketches the waveform of the voltage of capacitors C_1 and C_2. In order to allow the oscillator to have a pair of 50% duty cycle complementary outputs, the charging and discharging currents should be the same, so are the two integrating capacitors.

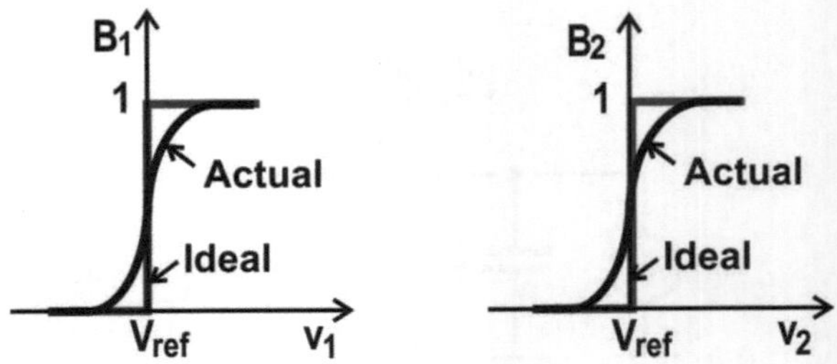

Fig. 4.16 Voltage transfer characteristics of comparators

The ideal and actual voltage transfer characteristics of the comparators are shown in Fig. 4.16. The actual voltage transfer characteristics of the comparators can be approximated using the following hyperbolic tangent function

$$B_n = \frac{1}{2}\left[\tanh(v_n - V_{ref}) + 1\right], \tag{4.46}$$

$n = 1, 2$. In the vicinity of the state transition of the comparators where $v_n - V_{ref}$ is small, $\tanh(v_n - V_{ref})$ can be approximated by its third-order Taylor series expansion

$$B_n \approx \frac{1}{2}\left[\left(v_n - V_{ref}\right) - \frac{1}{3}\left(v_n - V_{ref}\right)^3 + 1\right]. \tag{4.47}$$

Since voltage comparators have a finite transition slope, the hyperbolic tangent function imitates the operation of practical voltage comparators.

Since V_{ref} is constant, in order to simplify analysis, we shift the minimum voltage of the circuit in Fig. 4.14 from 0 to a negative constant voltage $-V_{DC}$ such as the reference voltage of the comparators is changed from V_{ref} to 0, as shown in Fig. 4.17. Equations (4.46) and (4.46) in this case are changed to

$$B_n = \frac{1}{2}\left[\tanh(v_n) + 1\right] \tag{4.48}$$

and

$$B_n \approx \frac{1}{2}\left(v_n - \frac{1}{3}v_n^3 + 1\right). \tag{4.49}$$

It is seen from (4.49) that the truncated Taylor series expansion of B_n is much simpler as compared with the one given in (4.47). Such a simplification is needed as it will greatly simplify analysis when we further expand v_n in its Volterra series expansion, as to be seen shortly.

The waveform of v_1 and v_2 of the relaxation oscillator of Fig. 4.17 with the threshold voltage of the comparators at 0 V are shown in Fig. 4.18. It is seen that the maximum voltage of v_1 and that of v_2 are changed from V_{ref} to 0 V.

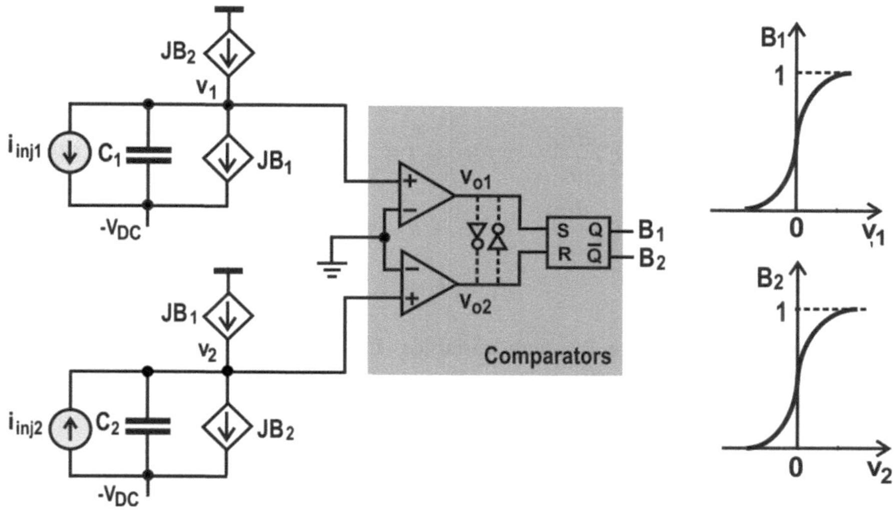

Fig. 4.17 Simplified schematic of dual-comparator relaxation oscillator. The minimum voltage is shifted from 0 to $-V_{DC}$ such as the threshold voltage of the comparators is changed from V_{ref} to 0

Fig. 4.18 Waveforms of the voltages of dual-comparator relaxation oscillator with threshold voltage of the comparators adjusted to 0 V

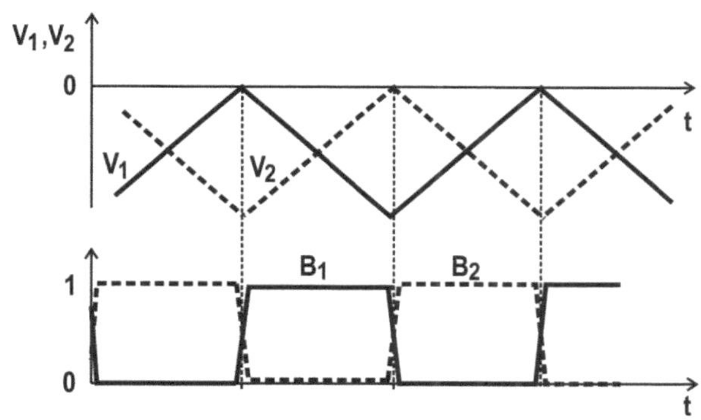

4.9 Volterra Circuits of Relaxation Oscillator

We showed in the previous sections that a nonlinear circuit can be represented by a set of Volterra circuits that are linear, have the same topology, element type, and element value but different inputs. We also showed that the voltage transfer characteristics of voltage comparators can be depicted using a hyperbolic tangent function, and in the vicinity of the state transition of the output of the comparators, the hyperbolic tangent function can be approximated by its truncated Taylor series expansion, typically to the order of 3. In this section, we derive the Volterra circuits of the dual-comparator relaxation oscillator using the approach introduced in Sect. 4.7.

Consider the dual-comparator relaxation oscillator with differential current injections shown in Fig. 4.17. Write Kirchhoff's Current Law at nodes 1 and 2 and note the complementary relation between the outputs of the SR latch

$$C_1\frac{dv_1}{dt} - 2JB_2 + J + i_{inj1} = 0, \tag{4.50a}$$

$$C_2\frac{dv_2}{dt} - 2JB_1 + J - i_{inj2} = 0, \tag{4.50b}$$

$$B_1 + B_2 = 1. \tag{4.50c}$$

Note that B_1 and B_2 are Boolean variables. Let the amplitude of the injection currents be changed from i_{inj} to ϵi_{inj}. Represent the resultant nodal voltage v_n of the oscillator in its third-order Volterra series expansion

$$v_n \approx \sum_{k=1}^{3} v_{n,k}\epsilon^k, \tag{4.51}$$

$n = 1, 2$. Substituting (4.51) into (4.49) gives

$$B_n \approx \frac{1}{2}+\frac{1}{2}\left(v_{n,1}\epsilon + v_{n,2}\epsilon^2+v_{n,3}\epsilon^3\right) - \frac{1}{6}\left(v_{n,1}\epsilon + v_{n,2}\epsilon^2+v_{n,3}\epsilon^3\right)^3, \tag{4.52}$$

$n = 1, 2$. Substituting (4.51) and (4.52) into (4.50) yields

$$\begin{aligned} &C_1\frac{d}{dt}\left(v_{1,1}\epsilon + v_{1,2}\epsilon^2 + v_{1,3}\epsilon^3\right) \\ &-2J\left[\frac{1}{2}+\frac{1}{2}\left(v_{2,1}\epsilon + v_{2,2}\epsilon^2 + v_{2,3}\epsilon^3\right) - \frac{1}{6}\left(v_{2,1}\epsilon + v_{2,2}\epsilon^2 + v_{2,3}\epsilon^3\right)^3\right] \\ &+J + i_{in1}\epsilon = 0, \end{aligned} \tag{4.53a}$$

$$\begin{aligned} &C_2\frac{d}{dt}\left(v_{2,1}\epsilon + v_{2,2}\epsilon^2 + v_{2,3}\epsilon^3\right) \\ &-2J\left[\frac{1}{2}+\frac{1}{2}\left(v_{1,1}\epsilon + v_{1,2}\epsilon^2 + v_{1,3}\epsilon^3\right) - \frac{1}{6}\left(v_{1,1}\epsilon + v_{1,2}\epsilon^2 + v_{1,3}\epsilon^3\right)^3\right] \\ &+J - i_{in2}\epsilon = 0. \end{aligned} \tag{4.53b}$$

$$\begin{aligned} &\frac{1}{2}+\frac{1}{2}\left(v_{1,1}\epsilon + v_{1,2}\epsilon^2 + v_{1,3}\epsilon^3\right) - \frac{1}{6}\left(v_{1,1}\epsilon + v_{1,2}\epsilon^2 + v_{1,3}\epsilon^3\right)^3 \\ &+\frac{1}{2}+\frac{1}{2}\left(v_{2,1}\epsilon + v_{2,2}\epsilon^2 + v_{2,3}\epsilon^3\right) - \frac{1}{6}\left(v_{2,1}\epsilon + v_{2,2}\epsilon^2 + v_{2,3}\epsilon^3\right)^3 = 0. \end{aligned} \tag{4.53c}$$

The three expressions in (4.53) are polynomials in ϵ. Since their value is 0 for arbitrary ϵ, the coefficient of the first-, second-, and third-order terms of these polynomials must also be 0. We therefore have

$$C_1\frac{dv_{1,k}}{dt} - Jv_{2,k} = -i_{in1,k}, \tag{4.54a}$$

$$C_2\frac{dv_{2,k}}{dt} - Jv_{1,k} = i_{in2,k}, \tag{4.54b}$$

$$v_{1,k} + v_{2,k} = 0, \tag{4.54c}$$

$k = 1, 2, 3$. The inputs of the first-, second-, and third-order Volterra circuit, denoted by $i_{in1,1}$ and $i_{in2,1}$, $i_{in1,2}$ and $i_{in2,2}$, and $i_{in1,3}$ and $i_{in2,3}$, respectively, are given by

$$i_{in1,1} = -i_{inj}, \tag{4.55a}$$

$$i_{in2,1} = i_{inj}. \tag{4.55b}$$

$$i_{in1,2} = 0, \tag{4.55c}$$

$$i_{in2,2} = 0, \tag{4.55d}$$

$$i_{in1,3} = \frac{J}{3}v_{2,1}^3, \tag{4.55e}$$

$$i_{in2,3} = \frac{J}{3}v_{1,1}^3. \tag{4.55f}$$

The preceding results reveal that the dual-comparator relaxation oscillator depicted by (4.50) can be represented by the Volterra circuits depicted by (4.54) with their inputs given by (4.55). These Volterra circuits are linear and identical except their inputs as the right-hand side of (4.54) is the same for all Volterra circuits while the left-hand side of (4.54) differs. The inputs of the first-order Volterra circuit are the actual injection currents of the oscillator. Since the inputs of second-order Volterra circuit are zero, its outputs $v_{1,2}$ and $v_{2,2}$ are zero as well. The inputs of the third-order Volterra are the nonlinear functions of the response of the first-order Volterra circuit and are readily available once the response of the first-order Volterra circuit is obtained. Once the responses of the first- and third-order Volterra circuits of the oscillator are available, the response of the oscillator can be obtained by summing up the responses of the first- and third-order Volterra circuits of the oscillator. As the Volterra circuits are linear circuits, they can be solved using methods for linear systems conveniently. Difficulties encountered in the analysis of injection-locked nonharmonic relaxation oscillators where harsh nonlinearities exist are averted.

The first-order Volterra circuit of the oscillator is governed by (4.54) with $k = 1$. The circuit implementation of the first and second expressions in (4.54) is straightforward while that of the third expression in (4.54) is rather not. The circuit

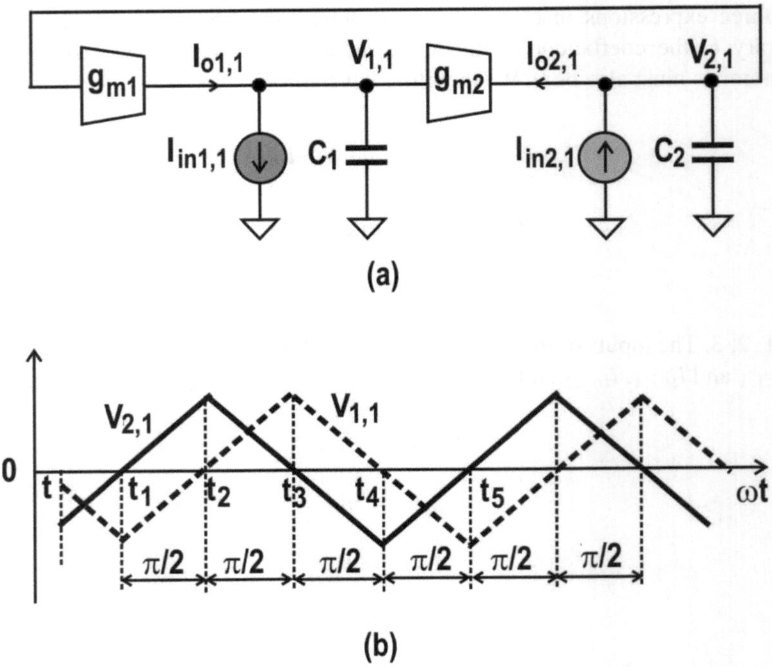

Fig. 4.19 Possible first-order Volterra circuit of dual-comparator relaxation oscillator. (**a**) Block diagram; (**b**) waveforms

shown in Fig. 4.19 might be one possible implementation of (4.54). Note that the direction of the output current of the two transconductors differs. Let us examine the circuit in detail. Consider a time instant $0 < t < t_1$ at which $v_{2,1} < 0$ and $v_{2,1}$ rises. Since $i_{o1,1}$ in this case flows away from C_1, $v_{1,1}$ drops, as shown in Fig. 4.19. At $t = t_1$ where $v_{2,1}$ just exceeds 0, $i_{o1,1}$ changes its direction and the current now flows into C_1. $v_{1,1}$ starts to rise. Since $v_{1,1} < 0$ when $t < t_1$, the direction of $i_{o2,1}$ remains unchanged and $v_{2,1}$ continues to rise. At $t = t_2$ where $v_{1,1}$ becomes positive, the direction of $i_{o2,1}$ changes and the current now flows away from C_2. As a result, $v_{2,1}$ starts to drop. Note since $v_{2,1} > 0$ for $t_2 < t < t_3$, $i_{o1,1}$ continues to charge C_1 and $v_{1,1}$ continues to climb. At $t = t_3$ where $v_{2,1} < 0$, the direction of $i_{o1,1}$ changes and $v_{1,1}$ starts to drop. The preceding process repeats and oscillation continues. It is observed that the phase difference between $v_{1,1}$ and $v_{2,1}$ is $\pi/2$. As a result, the third constraint of (4.54) is not satisfied. If either $v_{1,1}$ or $v_{2,1}$ is phase-shifted by $\pi/2$, the third constraint of (4.54) will be satisfied. It should be noted that the waveforms sketched in Fig. 4.19 assumes that the charging or discharging currents do not change in each half of the period.

It is seen from Fig. 4.19 that the sum of $v_{1,1}$ and $v_{2,1}$ is nonzero. As a result, the constraint imposed by the third expression in (4.54) is not satisfied. If additional phase-shift blocks are added, as shown in Fig. 4.20, then this constraint will be

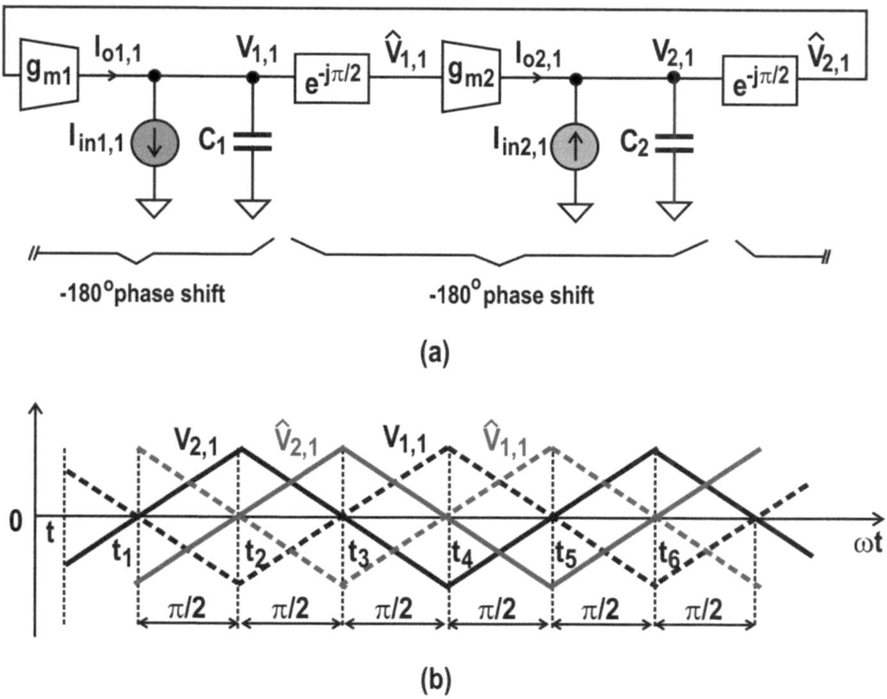

Fig. 4.20 First-order Volterra circuit of dual-comparator relaxation oscillator. (**a**) Block diagram; (**b**) waveforms

satisfied. To demonstrate this, consider the case where the injection currents are absent. Since

$$\begin{aligned} V_{1,1} &= \frac{1}{j\omega_o C_1} g_{m1} e^{-j\frac{\pi}{2}} V_{2,1} \\ &= \left(\frac{1}{\omega_o C_1} e^{-j\frac{\pi}{2}}\right) g_{m1} e^{-j\frac{\pi}{2}} V_{2,1}. \end{aligned} \tag{4.56}$$

Equation (4.56) shows that the phase shift from $V_{2,1}$ to $V_{1,1}$ is $-180°$. Similarly,

$$\begin{aligned} V_{2,1} &= \frac{1}{j\omega_o C_2} g_{m2} e^{-j\frac{\pi}{2}} V_{1,1} \\ &= \left(\frac{1}{\omega_o C_2} e^{-j\frac{\pi}{2}}\right) g_{m2} e^{-j\frac{\pi}{2}} V_{1,1}. \end{aligned} \tag{4.57}$$

The total phase shift from $V_{1,1}$ to $V_{2,1}$ is also $-180°$. As a result, the total phase shift around the loop is $-360°$. Barkhausen phase criterion is therefore satisfied.

To verify whether the constraint imposed by the third expression in (4.54) is not satisfied or not, consider a time instant $0 < t < t_1$ at which $v_{2,1} < 0$ and $v_{2,1}$ rises. Since $i_{o1,1}$ in this case flows away from C_1, $v_{1,1}$ drops. At $t = t_1$ where $v_{2,1}$ just exceeds 0, $i_{o1,1}$ would change its direction and the current would now flow into C_1 if the delay block was not present. The existence of the delay block will delay the threshold-crossing of the input of the downstream transconductor by a quarter of the period of the oscillator. As a result, when $\hat{V}_{2,1}$ exceeds the threshold-crossing voltage at t_2, $v_{1,1}$ will start to rise. Since $v_{1,1} < 0$ when $t > t_2$, $i_{o2,1}$ flows away from node 2 and $v_{2,1}$ starts to drop. At $t = t_3$ where $v_{2,1}$ drops below the threshold voltage, the existence of the delay block delays the threshold-crossing of $\hat{V}_{2,1}$ by a quarter of the period of the oscillator. As a result, $V_{2,1}$ continues to drop until t_4 at which $\hat{V}_{2,1}$ drops below the threshold. When this occurs, $V_{1,1}$ will start to drop. The preceding process repeats and oscillation is sustained. It is observed that the phase difference between $v_{1,1}$ and $v_{2,1}$ is now π. Since $g_{m1} = g_{m2}$ and $C_1 = C_2$, $V_{1,1}$ and $V_{2,1}$ have the same magnitude. As a result, the third constraint of (4.54) is satisfied. The addition of two delay blocks also ensures that $v_{1,1}$ and $v_{2,1}$ are out of phase.

The preceding analysis shows that the dual-comparator relaxation oscillator can be modeled using the system shown in Fig. 4.21 to account for both the nonlinear characteristics of the comparators and the constraint $B_1 + B_2 = 1$ of the SR latch. The nonlinear transconductors depicting the comparator are characterized by

$$i_{o1} = J\left(\hat{v}_2 - \frac{1}{3}\hat{v}_2^3\right), \tag{4.58a}$$

$$i_{o2} = J\left(\hat{v}_1 - \frac{1}{3}\hat{v}_1^3\right), \tag{4.58b}$$

where $\hat{v}_n$ is the delayed version of v_n.

When the amplitude of the injection currents is changed from i_{inj} to ϵi_{inj}, the governing equations of Fig. 4.21 will be given by

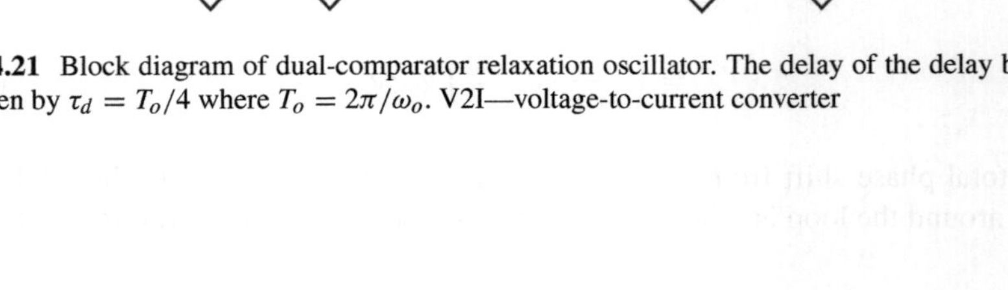

Fig. 4.21 Block diagram of dual-comparator relaxation oscillator. The delay of the delay blocks is given by $\tau_d = T_o/4$ where $T_o = 2\pi/\omega_o$. V2I—voltage-to-current converter

$$C_1 \frac{d}{dt}\left(v_{1,1}\epsilon + v_{1,2}\epsilon^2 + v_{1,3}\epsilon^3\right) \tag{4.59a}$$
$$-J\left[\left(\hat{v}_{2,1}\epsilon + \hat{v}_{2,2}\epsilon^2 + \hat{v}_{2,3}\epsilon^3\right) - \frac{1}{3}\left(\hat{v}_{2,1}\epsilon + \hat{v}_{2,2}\epsilon^2 + \hat{v}_{2,3}\epsilon^3\right)^3\right]$$
$$+i_{in1}\epsilon = 0,$$

$$C_2 \frac{d}{dt}\left(v_{2,1}\epsilon + v_{2,2}\epsilon^2 + v_{2,3}\epsilon^3\right) \tag{4.59b}$$
$$-J\left[\left(\hat{v}_{1,1}\epsilon + \hat{v}_{1,2}\epsilon^2 + \hat{v}_{1,3}\epsilon^3\right) - \frac{1}{3}\left(\hat{v}_{1,1}\epsilon + \hat{v}_{1,2}\epsilon^2 + \hat{v}_{1,3}\epsilon^3\right)^3\right]$$
$$-i_{in2}\epsilon = 0.$$

The expressions in (4.59) are polynomials in ϵ. Since their value is 0 for arbitrary ϵ, the coefficient of all the terms of the polynomials must also be 0. We therefore have

$$C_1 \frac{dv_{1,k}}{dt} - J\hat{v}_{2,k} + i_{in1,k} = 0, \tag{4.60a}$$

$$C_2 \frac{dv_{2,k}}{dt} - J\hat{v}_{1,k} - i_{in2,k} = 0, \tag{4.60b}$$

$k = 1, 2, 3$. The inputs of the first-, second-, and third-order Volterra circuit, denoted by $i_{in1,1}$ and $i_{in2,1}$, $i_{in1,2}$ and $i_{in2,2}$, and $i_{in1,3}$ and $i_{in2,3}$, respectively, are given by

$$i_{in1,1} = -i_{inj}, \tag{4.61a}$$

$$i_{in2,1} = i_{inj}. \tag{4.61b}$$

$$i_{in1,2} = 0, \tag{4.61c}$$

$$i_{in2,2} = 0, \tag{4.61d}$$

$$i_{in1,3} = \frac{J}{3}\hat{v}_{2,1}^3, \tag{4.61e}$$

$$i_{in2,3} = \frac{J}{3}\hat{v}_{1,1}^3. \tag{4.61f}$$

Figure 4.22 shows the block diagram of the first- and third-order Volterra circuits of the dual-comparator relaxation oscillator.

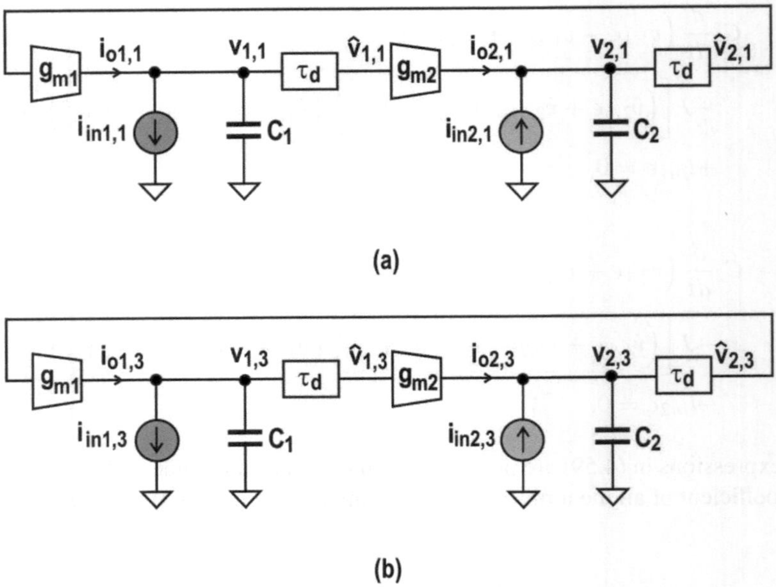

Fig. 4.22 Block diagram of the (**a**) first- and (**b**) third-order Volterra circuits of the dual-comparator relaxation oscillator

4.10 Effective Injection Signals

In this section, we derive the effective injection signal of the first-order Volterra circuit of the dual-comparator relaxation oscillator. One might argue that we already have the injection signals of the first-order Volterra circuit and they are given by the first two expressions of (4.60). These expressions are indeed part of but not the complete injection signals of the first-order Volterra circuit. We will show in this section that the high-order Volterra circuits of the oscillator contribute to the injection signals of the first-order Volterra circuit at ω_{inj}. The contribution of the high-order Volterra circuits changes the total or effective injection signals of the first-order Volterra circuit.

The block diagram of the first-order Volterra circuit of the oscillator is given in Fig. 4.22. Let the phase of $V_{2,1}$ be chosen to be the reference phase. The phase of $V_{1,1}$ is defined with reference to $V_{2,1}$. Let the external differential injection current be

$$i_{in2,1} = I_{inj}\sin(\omega_{inj}t + \phi_{inj}), \tag{4.62a}$$

$$i_{in1,1} = -I_{inj}\sin(\omega_{inj}t + \phi_{inj}), \tag{4.62b}$$

where ϕ_{inj} is the injection phase of $i_{in2,1}$. Assume that the frequency of the injection signal $\omega_{inj} = \omega_o + \Delta\omega$ falls into the lock range of the oscillator such that the oscillator will lock to the injection signal and oscillate at ω_{inj} once it is locked to the injection signal. The oscillator is depicted by its first-, second-, and third-order Volterra circuits. Since the first-order Volterra circuit is linear with its inputs at ω_{inj}. The outputs of the circuit only contain frequency components at ω_{inj} and can be written as

$$v_{2,1} = V_{2,1,m} \sin\left(\omega_{inj} t\right), \tag{4.63a}$$

$$v_{1,1} = -V_{1,1,m} \sin\left(\omega_{inj} t\right). \tag{4.63b}$$

Note that the phase difference between $v_{1,1}$ and $v_{2,1}$ is 180°. The delayed version of $v_{2,1}$ and $v_{1,1}$ is given by

$$\hat{v}_{2,1} = V_{2,1,m} \sin\left(\omega_{inj} t - \frac{\pi}{2}\right), \tag{4.64a}$$

$$\hat{v}_{1,1} = -V_{1,1,m} \sin\left(\omega_{inj} t - \frac{\pi}{2}\right). \tag{4.64b}$$

To obtain the inputs of the third-order Volterra circuit, since

$$\hat{v}^3_{2,1} = \frac{3}{4} V^3_{2,1,m} \sin\left(\omega_{inj} t - \frac{\pi}{2}\right) - \frac{1}{4} V^3_{2,1,m} \sin\left(3\omega_{inj} t - \frac{3\pi}{2}\right), \tag{4.65a}$$

$$\hat{v}^3_{1,1} = -\frac{3}{4} V^3_{1,1,m} \sin\left(\omega_{inj} t - \frac{\pi}{2}\right) + \frac{1}{4} V^3_{1,1,m} \sin\left(3\omega_{inj} t - \frac{3\pi}{2}\right), \tag{4.65b}$$

we arrive at the input of the third-order Volterra circuit from (4.60)

$$i_{in1,3} = \frac{J}{4} V^3_{2,1,m} \sin\left(\omega_{inj} t - \frac{\pi}{2}\right) - \frac{J}{12} V^3_{2,1,m} \sin\left(3\omega_{inj} t - \frac{3\pi}{2}\right), \tag{4.66a}$$

$$i_{in2,3} = -\frac{J}{4} V^3_{1,1,m} \sin\left(\omega_{inj} t - \frac{\pi}{2}\right) + \frac{J}{12} V^3_{1,1,m} \sin\left(3\omega_{inj} t - \frac{3\pi}{2}\right). \tag{4.66b}$$

Equation (4.66) shows that the inputs of the third-order Volterra circuit contain two frequency components, one at ω_{inj} and the other at $3\omega_{inj}$. As a result, the output of the circuit also has only two frequency components, one at ω_{inj} and the other at $3\omega_{inj}$.

Since the first- and third-order Volterra circuits are identical except their inputs, they have the same transfer function. We can therefore combine the input of the first-order Volterra circuit and that of the third-order Volterra circuit, both at ω_{inj}, to obtain the total or effective injection signals of the first-order Volterra circuit at ω_{inj}

$$I_{in1,1,eff} = -I_{inj}\sin\left(\omega_{inj}t + \phi_{inj}\right) + \frac{J}{4}V_{2,1,m}^3 \sin\left(\omega_{inj}t - \frac{\pi}{2}\right)$$
$$= -I_{inj}\sin\left(\omega_{inj}t + \phi_{inj}\right) - \frac{J}{4}V_{2,1,m}^3 \cos\left(\omega_{inj}t\right), \tag{4.67a}$$

$$I_{in2,1,eff} = I_{inj}\sin\left(\omega_{inj}t + \phi_{inj}\right) - \frac{J}{4}V_{1,1,m}^3 \sin\left(\omega_{inj}t - \frac{\pi}{2}\right)$$
$$= I_{inj}\sin\left(\omega_{inj}t + \phi_{inj}\right) + \frac{J}{4}V_{1,1,m}^3 \cos\left(\omega_{inj}t\right). \tag{4.67b}$$

In order to maximize the effective injection signals of the first-order Volterra circuit, the components contributed by the third-order Volterra circuit should be in phase with the external injection signals. To achieve this, a natural choice of the injection phase of the external injection signal at node 2 is

$$\phi_{inj}^* = \frac{\pi}{2}. \tag{4.68}$$

The maximum effective injection signals at the optimal injection phase $\phi_{inj}^* = \frac{\pi}{2}$ are obtained from (4.67)

$$I_{in1,1,eff,max} = -\left(I_{inj} + \frac{J}{4}V_{2,1,m}^3\right)\cos\left(\omega_{inj}t\right), \tag{4.69a}$$

$$I_{in2,1,eff,max} = \left(I_{inj} + \frac{J}{4}V_{1,1,m}^3\right)\cos\left(\omega_{inj}t\right). \tag{4.69b}$$

It is seen from (4.69) that the contribution of the third-order Volterra circuit increases the effective injection signals of the first-order Volterra circuit beyond the external injection signals. As the lock range of a harmonic oscillator is proportional to the strength of its injection signals, strengthening the effective injection signals of the first-order Volterra circuit, accredited to the contribution of the third-order Volterra circuit, results in a larger lock range of the first-order Volterra circuit. Because the frequency of the first-order Volterra circuit of the relaxation oscillator is the frequency of the first harmonic of the oscillator, the maximum frequency shift of the first-order Volterra circuit caused by the effective injection signals is the same as that of the relaxation oscillator and therefore is the lock range of the relaxation oscillator.

It is interesting to note that since the voltage across C_2 generated by the external injection signal i_{inj2} is given by

$$\frac{1}{j\omega C_2}I_{inj2}e^{j\phi_{inj}} = \frac{I_{inj2}}{\omega C_2}e^{j\left(\phi_{inj}-\frac{\pi}{2}\right)}$$

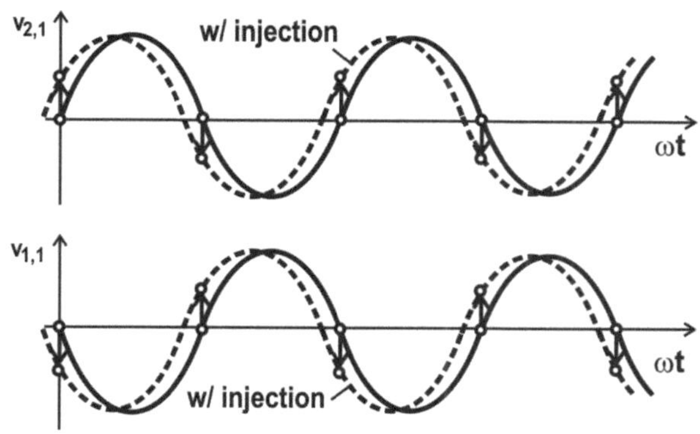

Fig. 4.23 Optimal injection phase occurs when the phase of the voltage of C_2 induced by i_{inj2} is in phase with $v_{2,1}$, and the phase of the voltage of C_1 induced by i_{inj1} is in phase with $v_{1,1}$

and the phase of $v_{2,1}$ is 0, the impact of i_{inj2} on the frequency of the oscillator is maximized if the phase of the above quantity is 0. We therefore have $\phi_{inj} = \frac{\pi}{2}$. This is the case shown in Fig. 4.23.

4.11 Lock Range

With the availability of the effective injection signals of the first-order Volterra circuit, we are now in the position to derive the lock range of the dual-comparator relaxation oscillator.

Write the governing equations of the first-order Volterra circuit derived earlier in the frequency domain with the external injection signals now replaced with the effective injection signals

$$sC_1V_{1,1} - JV_{2,1}e^{-j\frac{\pi}{2}} + I_{in1,1,eff} = 0, \tag{4.70}$$

$$sC_2V_{2,1} - JV_{1,1}e^{-j\frac{\pi}{2}} + I_{in2,1,eff} = 0. \tag{4.71}$$

Solve for $V_{1,1}$ and $V_{2,1}$

$$V_{1,1} = -\frac{sC_2I_{in1,1,eff} + Je^{-j\frac{\pi}{2}}I_{in2,1,eff}}{s^2C_1C_2 + J^2}, \tag{4.72a}$$

$$V_{2,1} = -\frac{sC_1I_{in2,1,eff} + Je^{-j\frac{\pi}{2}}I_{in1,1,eff}}{s^2C_1C_2 + J^2}. \tag{4.72b}$$

It was shown earlier that when the injection phase is at its optimal value $\phi^*_{inj} = \frac{\pi}{2}$, the effective injection signals of the first-order Volterra circuit will be maximized and the phase difference between the effective injection signals in this case will be 180°. We write the maximum effective injection signals in their phasor forms

$$I_{in1,1,eff} = -I_{eff1}, \tag{4.73a}$$

$$I_{in2,1,eff} = I_{eff1}, \tag{4.73b}$$

where I_{eff1} is the magnitude of $I_{in1,1,eff}$ and $I_{in2,1,eff}$, with the understanding that the phasors in (4.73) are at frequency ω_{inj}. Substituting (4.73) into (4.72) yields

$$\begin{aligned} V_{1,1} &= \frac{I_{eff1}}{J}\frac{1}{1-\frac{\omega_{inj}}{\omega_o}}e^{j\frac{\pi}{2}} \\ &= -\frac{I_{eff1}}{J}\left(\frac{\omega_o}{\Delta\omega}\right)e^{j\frac{\pi}{2}}, \end{aligned} \tag{4.74a}$$

$$\begin{aligned} V_{2,1} &= \frac{I_{eff1}}{J}\frac{1}{1-\frac{\omega_{inj}}{\omega_o}}e^{-j\frac{\pi}{2}} \\ &= -\frac{I_{eff1}}{J}\left(\frac{\omega_o}{\Delta\omega}\right)e^{-j\frac{\pi}{2}}. \end{aligned} \tag{4.74b}$$

It is seen from (4.74) that the difference between the phase of $V_{1,1}$ and that of $V_{2,1}$ is 180°. Also observed is that $V_{1,1}$ and $V_{2,1}$ have the same magnitude. Define

$$H_{o1} = \frac{J}{sC_1}, \tag{4.75a}$$

$$H_{o2} = \frac{J}{sC_2}, \tag{4.75b}$$

$$H_o = H_{o1}H_{o2}. \tag{4.76}$$

Since when the injection signals are absent, the oscillator will oscillate at ω_o. Barkhausen criteria must therefore be satisfied at ω_o, i.e.,

$$1 + H_{o1}(\omega_o)H_{o2}(\omega_o) = 0. \tag{4.77}$$

With $C_1 = C_2 = C$, we have from (4.77)

$$\frac{J}{\omega C} = 1. \tag{4.78}$$

As a result,

$$H_{o1}(\omega_o),\, H_{o2}(\omega_o) = e^{-j\frac{\pi}{2}}. \tag{4.79}$$

Making use of (4.75), we can write (4.72) as

$$V_{1,1} = -\frac{1}{J}\frac{H_o}{1+H_o}\left(\frac{1}{H_{o2}}I_{in1,1,eff} + e^{-j\frac{\pi}{2}}I_{in2,1,eff}\right), \tag{4.80a}$$

$$V_{2,1} = -\frac{1}{J}\frac{H_o}{1+H_o}\left(\frac{1}{H_{o1}}I_{in2,1,eff} + e^{-j\frac{\pi}{2}}I_{in1,1,eff}\right). \tag{4.80b}$$

Note that all variables in (4.80) are evaluated at ω_{inj}. Since $\omega_{inj} = \omega_o + \Delta\omega$ with $\Delta\omega \ll \omega_o$, $H_o(\omega_o + \Delta\omega)$ can be approximated with its first-order Taylor series expansion

$$H_o(\omega_o + \Delta\omega) \approx H_o(\omega_o) + \left(\frac{dH_o}{d\omega}\right)_{\omega_o}\Delta\omega. \tag{4.81}$$

Further utilizing

$$H_o(\omega_o + \Delta\omega) \approx H_o(\omega_o) = -1, \tag{4.82}$$

we have from (4.80)

$$V_{1,1} \approx \frac{1}{J}\frac{1}{\left(\frac{dH_o}{d\omega}\right)_{\omega_o}\Delta\omega}\left(\frac{1}{H_{o2}}I_{in1,1,eff} + e^{-j\frac{\pi}{2}}I_{in2,1,eff}\right), \tag{4.83a}$$

$$V_{2,1} \approx \frac{1}{J}\frac{1}{\left(\frac{dH_o}{d\omega}\right)_{\omega_o}\Delta\omega}\left(\frac{1}{H_{o1}}I_{in1,1,eff} + e^{-j\frac{\pi}{2}}I_{in2,1,eff}\right). \tag{4.83b}$$

It follows from (4.83) that

$$V_{1,1}^* \approx \frac{1}{J}\frac{1}{\left(\frac{dH_o^*}{d\omega}\right)_{\omega_o}\Delta\omega}\left(\frac{1}{H_{o2}^*}I_{in1,1,eff}^* + e^{j\frac{\pi}{2}}I_{in2,1,eff}^*\right), \tag{4.84a}$$

$$V_{2,1}^* \approx \frac{1}{J}\frac{1}{\left(\frac{dH_o^*}{d\omega}\right)_{\omega_o}\Delta\omega}\left(\frac{1}{H_{o1}^*}I_{in1,1,eff}^* + e^{j\frac{\pi}{2}}I_{in2,1,eff}^*\right). \tag{4.84b}$$

We therefore have

$$|V_{1,1}|^2 \approx \frac{1}{J^2} \frac{1}{\left(\frac{dH_o}{d\omega}\frac{dH_o^*}{d\omega}\right)_{\omega_o} (\Delta\omega)^2} \times \left(|I_{in2,1,eff}|^2 + \frac{|I_{in1,1,eff}|^2}{|H_{o2}|^2} + \frac{e^{j\frac{\pi}{2}}}{H_{o2}} I_{in1,1,eff} I^*_{in2,1,eff} + \frac{e^{-j\frac{\pi}{2}}}{H^*_{o2}} I^*_{in1,1,eff} I_{in2,1,eff} \right). \tag{4.85}$$

Making use of the quality factor introduced earlier in (3.32), which is repeated here for convenience

$$\left(\frac{dH_o}{d\omega}\frac{dH_o^*}{d\omega}\right)_{\omega_o} = \left[\frac{2Q(\omega_o)}{\omega_o}\right]^2, \tag{4.86}$$

we have from (4.85)

$$|V_{1,1}|^2 \approx \frac{1}{J^2}\left[\frac{\omega_o}{2Q(\omega_o)}\right]^2 \frac{1}{(\Delta\omega)^2} \times \left(|I_{in2,1,eff}|^2 + \frac{|I_{in1,1,eff}|^2}{|H_{o2}|^2} + \frac{e^{j\frac{\pi}{2}}}{H_{o2}} I_{in1,1,eff} I^*_{in2,1,eff} + \frac{e^{-j\frac{\pi}{2}}}{H^*_{o2}} I^*_{in1,1,eff} I_{in2,1,eff} \right). \tag{4.87}$$

Solve for $\Delta\omega$ from (4.87)

$$|\Delta\omega| \approx \frac{1}{J}\frac{\omega_o}{2Q(\omega_o)}\frac{1}{|V_{1,1}|} \times \sqrt{|I_{in2,1,eff}|^2 + \frac{|I_{in1,1,eff}|^2}{|H_{o2}|^2} + \frac{e^{j\frac{\pi}{2}}}{H_{o2}} I_{in1,1,eff} I^*_{in2,1,eff} + \frac{e^{-j\frac{\pi}{2}}}{H^*_{o2}} I^*_{in1,1,eff} I_{in2,1,eff}}. \tag{4.88}$$

Making use of

$$H_{o1}(\omega_o + \Delta\omega) \approx H_{o1}(\omega_o) = e^{-j\frac{\pi}{2}}, \tag{4.89a}$$

$$H_{o2}(\omega_o + \Delta\omega) \approx H_{o2}(\omega_o) = e^{-j\frac{\pi}{2}}, \tag{4.89b}$$

which was derived earlier in (4.79) and noting that when the injection phase is $\frac{\pi}{2}$, the effective injection signals are maximized and their phase difference is 180°, i.e., $I_{in1,1,eff} = -I_{eff1}$ and $I_{in2,1,eff} = I_{eff1}$, we have

$$
\begin{aligned}
|\Delta\omega| &\approx \frac{1}{J}\frac{\omega_o}{2Q(\omega_o)}\frac{1}{|V_{1,1}|}\sqrt{I_{eff1}^2 + I_{eff1}^2 - e^{j\pi} I_{eff1}^2 - e^{-j\pi} I_{eff1}} \\
&\approx \frac{1}{J}\frac{\omega_o}{Q(\omega_o)}\frac{I_{eff1}}{|V_{1,1}|} \\
&= \frac{\omega_o}{Q(\omega_o)}\frac{I_{eff1}}{|I_{o2,1}|}.
\end{aligned}
\tag{4.90}
$$

Equation (4.90) is obtained using $V_{1,1}$. Since $V_{1,1}$ and $V_{2,1}$ are symmetrical, one can also use $V_{2,1}$ to derive the lock range. The result will be the same. It is seen from (4.90) that the lock range of the first-order Volterra circuit has the same expression as that of harmonic oscillator derived earlier. This is expected as the first-order Volterra circuit is also a harmonic oscillator. Since the frequency of the first-order Volterra circuit is the same as that of the first harmonic of the output of the oscillator, the lock range of the first-order Volterra circuit is the same as that of the oscillator.

It is also seen from (4.90) that the lock range is directly proportional to the effective injection signals of the first-order Volterra circuit. The larger the effective injection signals of the first-order Volterra circuit, the larger the lock range. As the effective injection signals of the first-order Volterra circuit are made of both the external injection signals and the contribution of the third-order Volterra circuit, one can increase the effective injection signals of the first-order Volterra circuit by increasing the contribution of the third-order Volterra circuit without increasing the external injection signals. Techniques on how to achieve this will be presented in the next chapter.

4.12 Chapter Summary

The oscillation of a three-stage ring oscillator was used to demonstrate how Barkhausen criteria can be used to study the oscillation of nonharmonic oscillators. We showed that the internal mechanism of the ring oscillator that enables the oscillator to automatically satisfy Barkhausen magnitude criterion is the input-modulated nonlinear transconductance of the delay stages of the oscillator. Barkhausen phase criterion is satisfied by the number of the delay stages of oscillators.

Relaxation oscillators distinct themselves from ring oscillators by the inclusion of comparators that have a harsher nonlinear voltage transfer characteristic as compared with the delay stages of ring oscillators. Similar to ring oscillators, the internal mechanism of relaxation oscillators that enables the oscillators to satisfy Barkhausen magnitude criterion is the input-modulated nonlinear transconductance of the transconductors that depict the comparators of the oscillator except that the degree of the nonlinearity of the transconductors of relaxation oscillators is higher as compared with those of ring oscillators. Barkhausen phase criterion is satisfied by the change of the direction of the current of the transconductors once the input voltage of the comparators crosses the threshold voltage of the comparators.

Since the waveform of the output of a nonharmonic oscillator is a square wave ideally, which contains an infinite number of harmonically related frequency components, the nonharmonic oscillator can be represented by a set of harmonic oscillators whose frequency is the first, third, and fifth harmonics of the output of the nonharmonic oscillator. The beauty of this approach is that it allows the use of linear system approaches valid for analyzing harmonic oscillators and the known knowledge of harmonic oscillators to explore the unknowns of nonharmonic oscillators. To investigate the theoretical basis upon which a nonharmonic oscillator can be represented by a set of harmonic oscillators whose frequencies are harmonically related, we first reviewed the fundamentals of Volterra series. Two representative nonlinearities encountered in integrated circuits, specifically, the nonlinear channel current of MOS transistors and the nonlinear capacitance of pn-junctions, were used as examples to demonstrate how a nonlinear element with mildly nonlinear characteristics can be represented by the truncated Taylor series expansion often to the order of 3 of the governing equation of the element. We further showed that the Volterra circuits of a nonlinear circuit can be obtained by first representing the variables of circuits, often nodal voltages and branch currents, in their truncated Volterra series expansion, and then perturbing the amplitude of the input of the circuit slightly such that the convergence of the Volterra series expansion of the variables of the circuits is not affected. The resultant variables of the circuit are the polynomials of the perturbation variable. The governing equations of the circuit with the consideration of the perturbed input are also the polynomials of the perturbation variable from which the governing equations of Volterra elements and Volterra circuits are obtained. Moreover, we showed that the Volterra circuits of a nonlinear circuit are linear, have the same topology, element type, and element value subsequently the same transfer function but different inputs. The inputs of the first-order Volterra circuit of the nonlinear circuit are the actual inputs of the nonlinear circuit while those of high-order Volterra circuits are due to the nonlinearity of the oscillator and are the nonlinear functions of the response of lower-order Volterra circuits. The actual inputs of the nonlinear circuit do not make their appearance in the inputs of Volterra circuits whose order is 2 and higher.

The Volterra series approach was used to analyze an injection-locked dual-comparator relaxation oscillator with a differential unitone sinusoidal injection signal. We showed that the relaxation oscillator can be represented by a set of identical harmonic oscillators whose inputs are harmonically related. We further showed that high-order Volterra circuits contribute to the effective injection signal of lower-order Volterra circuits. In the dual-comparator relaxation oscillator studied, the effective injection signal of the first-order Volterra circuit of the oscillator consists of two distinct parts: The external injection signals of the relaxation oscillator and the contribution of the third-order Volterra circuit, both at the frequency of the external injection signals. It is the contribution of the third-order Volterra circuit that strengthens the effective injection signal. When the external injection signals are injected at the threshold-crossings of the output of the relaxation oscillator, the effective injection signals are maximized so is the lock range of the relaxation oscillator. As the frequency of the first-order Volterra circuit is the same as that of the first harmonic of the dual-comparator relaxation oscillator, the lock range of the relaxation oscillator is the same as that of the first-order Volterra circuit.

Chapter 5
Injection-Locking Techniques for Nonharmonic Oscillators

5.1 Introduction

Lock range is one of the key performance indicators of injection-locked nonharmonic oscillators. A large lock range is critical for applications such as the ILFD pre-scalar of frequency synthesizers in digital systems. This is because not only the frequency of the injection-locked ring oscillators varies with PVT uncertainty, the injection-locking signal, which is the output of the ring oscillator of the synthesizer, is also affected by PVT uncertainty. The success of the injection-locked frequency dividers critically depends upon the lock range of the ring oscillators. A large lock range is also pivotal to the injection-lock based frequency calibration of PWMs. The system clock of these microsystems is usually generated locally using either a ring oscillator or a relaxation oscillator to take the advantages of their low power consumption and a small silicon area [104]. The frequency of the system clock of these systems is sensitive to PVT uncertainty. It needs to be calibrated against a reference clock of a constant frequency prior to their intended operations. As lock range is essential to the success of injection-lock based frequency calibration, techniques that increase the lock range of nonharmonic oscillators are critical.

Although the lock range of nonharmonic oscillators can always be increased by boosting the strength of external injection signals, increasing the strength of external injection signals is constrained by the maximum amount of the power that the transmitter is permitted to transmit. Also, since the frequency of the envelope of an ASK-modulated carrier is significantly lower as compared with that of the carrier, envelope-based frequency calibration is far more power-efficient as compared with carrier-based frequency calibration [73]. A small modulation index in this case is desired from a power harvest point of view as the smaller the modulation index, the less fluctuation the power flow from the transmitter to the PWM. Clearly in these applications, techniques that improve lock range without increasing the strength of injection signals are critical.

F. Yuan, *Injection-Locking in Mixed-Mode Signal Processing*,
https://doi.org/10.1007/978-3-030-17364-7_5

This chapter deals with techniques that increase the lock range of nonharmonic oscillators without increasing the strength of external injection signals. Section 5.2 studies the impact of the degree of the nonlinear characteristics of nonharmonic oscillators on the lock range of these oscillators. Section 5.3 investigates the lock range of dual-comparator relaxation oscillators with single unitone injection and that with dual unitone injections. In Sect. 5.4, the lock range of dual-comparator relaxation oscillators with dual multitone injections is derived. Section 5.5 compares the lock range of dual-comparator relaxation oscillators with single multitone injection and that with dual multitone injections. The impact of the duty cycle of multitone injection signals on the lock range of nonharmonic oscillators is investigated in Sect. 5.6. Section 5.7 focuses on the impact of the frequency attenuation mechanism of nonharmonic oscillators on the lock range of these oscillators. For the purpose of comparison, the impact of the frequency attenuation mechanism on the lock range of harmonic oscillators is also explored. Finally, the chapter is summarized in Sect. 5.8.

5.2 Degree of Nonlinearity

For ring oscillators, the nonlinear elements are the nonlinear transconductors depicting the delay stages of the oscillators. For the dual-comparator relaxation oscillator studied earlier, the nonlinear elements are the nonlinear transconductors modeling the comparators. We showed in Chap. 4 that the high-order Volterra circuits of nonharmonic oscillators contribute to the effective injection signals of the first-order Volterra circuit of the oscillators. In this section, we investigate how the degree of the nonlinear elements of the oscillators affects the effective injection signals of the first-order Volterra circuit of the oscillators subsequently the lock range of the oscillators. We use the dual-comparator relaxation oscillator studied earlier as an example to carry out the analysis. The underlined principle can be applied to other types of nonharmonic oscillators as well.

5.2.1 Nonlinearity Factor

In Chap. 4, we showed that the comparators of the dual-comparator relaxation oscillator can be depicted using a hyperbolic tangent function and in the vicinity of state transitions, the hyperbolic tangent function can be approximated with its third-order Taylor series expansion. In order to investigate the impact of the degree of the nonlinear characteristics of the comparators on the effective injection signals of the first-order Volterra circuit of the oscillator, nonlinearity factor a that changes the degree of the nonlinearity of the comparators is introduced and the comparators in this case are depicted by

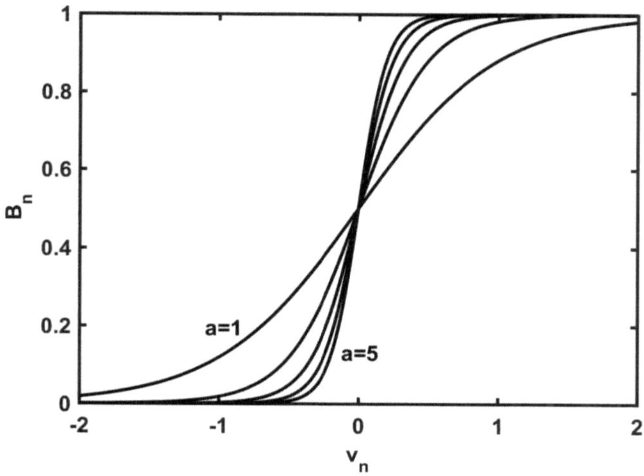

Fig. 5.1 Impact of nonlinearity factor a on the degree of the nonlinearity of the voltage transfer characteristics of comparators. a is varied from 1 to 5 with step size 1

$$B_n = \frac{1}{2}\left[\tanh\left(av_n\right) + 1\right], \tag{5.1}$$

where $n = 1, 2$. Varying a alters the degree of the nonlinearity of the comparators, as illustrated graphically in Fig. 5.1. It is seen that increasing a hardens the nonlinear characteristics of the comparators.

In the vicinity of the state transition of the comparators, $\tanh(av_n)$ can be approximated by its third-order Taylor series expansion

$$B_n \approx \frac{1}{2}\left[av_n - \frac{1}{3}\left(av_n\right)^3 + 1\right]. \tag{5.2}$$

Figure 5.2 compares the hyperbolic tangent function given by (5.1) and its third-order Taylor series expansion given by (5.2). It is seen that they agree well when v_n is small and their difference becomes noticeable once v_n moves away from 0. Also observed is that the larger a, the smaller the region in which the difference between (5.1) and (5.2) is small.

5.2.2 Volterra Circuits

Consider the dual-comparator relaxation oscillator with differential current injections shown in Fig. 4.17. The governing equations of the oscillator given in (4.50) were obtained from Kirchhoff's Current Law at nodes 1 and 2 while accounting for

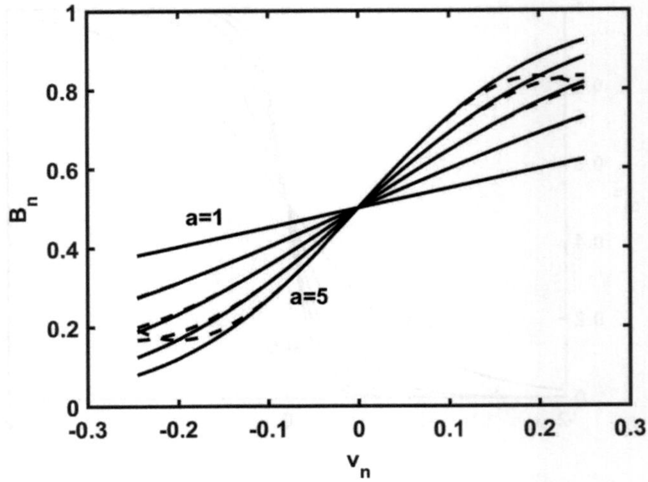

Fig. 5.2 Comparison of hyperbolic tangent function (5.1) and its third-order Taylor series expansion (5.2). a is varied from 1 to 5 with step size 1. Legends: Solid line: $B_n = \frac{1}{2}\left[\tanh\left(av_n\right)+1\right]$. Dashed line: $B_n \approx \frac{1}{2}\left[av_n - \frac{1}{3}\left(av_n\right)^3 + 1\right]$

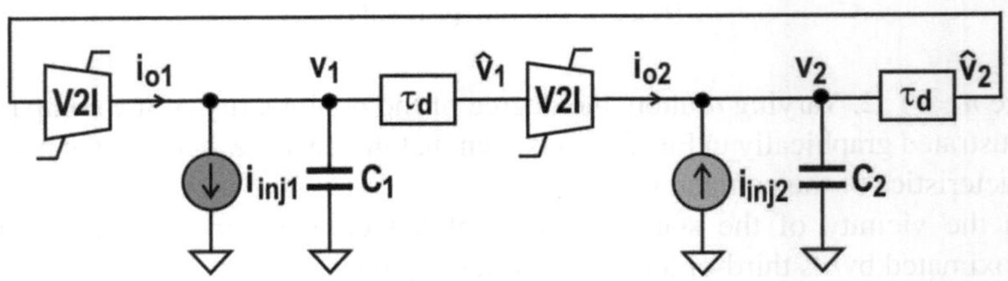

Fig. 5.3 Block diagram of dual-comparator relaxation oscillator. $\tau_d = T_o/4$ where $T_o = 2\pi/\omega_o$

the relation between the complementary outputs of the SR latch, and are repeated here for convenience

$$C_1\frac{dv_1}{dt} - 2JB_2 + J + i_{inj1} = 0, \tag{5.3a}$$

$$C_2\frac{dv_2}{dt} - 2JB_1 + J + i_{inj2} = 0, \tag{5.3b}$$

$$B_1 + B_2 = 1. \tag{5.3c}$$

It was shown in Sect. 4.9 that the third constraint in (5.3) can be accounted for by the insertion of two delay blocks shown in Fig. 4.21. For the sake of convenience, Fig. 4.21 is redrawn in Fig. 5.3. The degree of the nonlinearity of the transconductors, i.e., V2I blocks, is now modified with the inclusion of nonlinearity factor a.

Let the amplitude of the injection currents of the oscillator be changed from i_{inj} to ϵi_{inj} where ϵ is a constant. Represent the resultant nodal voltage v_n of the oscillator in its third-order Volterra series expansion and substitute the result into (5.2)

$$B_n \approx \frac{1}{2} + \frac{a}{2}\left(\sum_{k=1}^{3} v_{n,k}\epsilon^k\right) - \frac{a^3}{6}\left(\sum_{k=1}^{3} v_{n,k}\epsilon^k\right)^3. \tag{5.4}$$

Making use of the third-order Volterra series expansion of v_n and substituting (5.4) into (5.3) with the understanding that the constraint given in the third expression in (5.3) has been taken care of by the addition of the two delay blocks in Fig. 5.3

$$\begin{aligned}&C_1\frac{d}{dt}\left(v_{1,1}\epsilon + v_{1,2}\epsilon^2 + v_{1,3}\epsilon^3\right)\\&-J\left[a\left(\hat{v}_{2,1}\epsilon + \hat{v}_{2,2}\epsilon^2 + \hat{v}_{2,3}\epsilon^3\right) - \frac{a^3}{3}\left(\hat{v}_{2,1}\epsilon + \hat{v}_{2,2}\epsilon^2 + \hat{v}_{2,3}\epsilon^3\right)^3\right]\\&+i_{in1}\epsilon = 0,\end{aligned} \tag{5.5a}$$

$$\begin{aligned}&C_2\frac{d}{dt}\left(v_{2,1}\epsilon + v_{2,2}\epsilon^2 + v_{2,3}\epsilon^3\right)\\&-J\left[a\left(\hat{v}_{1,1}\epsilon + \hat{v}_{1,2}\epsilon^2 + \hat{v}_{1,3}\epsilon^3\right) - \frac{a^3}{3}\left(\hat{v}_{1,1}\epsilon + \hat{v}_{1,2}\epsilon^2 + \hat{v}_{1,3}\epsilon^3\right)^3\right]\\&-i_{in2}\epsilon = 0.\end{aligned} \tag{5.5b}$$

The expressions in (5.5) are polynomials in ϵ. Since their value is 0 for arbitrary ϵ, the coefficient of all the terms of the polynomials must also be 0. We therefore have

$$C_1\frac{dv_{1,k}}{dt} - aJ\hat{v}_{2,k} + i_{in1,k} = 0, \tag{5.6a}$$

$$C_2\frac{dv_{2,k}}{dt} - aJ\hat{v}_{1,k} - i_{in2,k} = 0, \tag{5.6b}$$

$k = 1, 2, 3$. The inputs of the first-, second-, and third-order Volterra circuit, denoted by $i_{in1,1}$ and $i_{in2,1}$, $i_{in1,2}$ and $i_{in2,2}$, and $i_{in1,3}$ and $i_{in2,3}$, respectively, are given by

$$i_{in1,1} = -i_{inj}, \tag{5.7a}$$

$$i_{in2,1} = i_{inj}. \tag{5.7b}$$

$$i_{in1,2} = 0, \tag{5.7c}$$

$$i_{in2,2} = 0, \tag{5.7d}$$

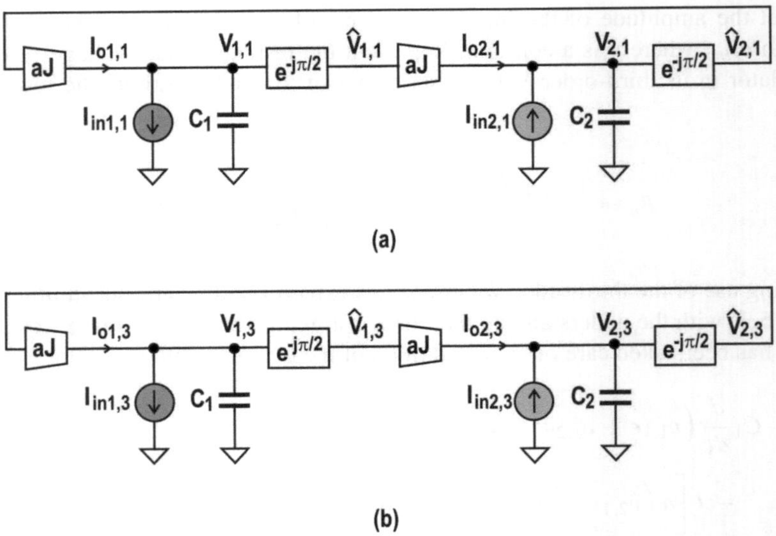

Fig. 5.4 (**a**) First- and (**b**) third-order Volterra circuits of dual-comparator relaxation oscillator with nonlinearity factor a to account for the degree of the nonlinearity of the transconductors depicting the comparators

$$i_{in1,3} = \frac{a^3 J}{3}\hat{v}_{2,1}^3, \tag{5.7e}$$

$$i_{in2,3} = \frac{a^3 J}{3}\hat{v}_{1,1}^3. \tag{5.7f}$$

A comparison of (5.6) and (5.7) with (4.54) and (4.55) reveals that nonlinearity factor a affects the governing equations of both the first- and third-order Volterra circuits. For the first-order Volterra circuit, it changes the transconductance of the transconductors from J to aJ, as shown in Fig. 5.4. For the third-order Volterra circuit, its impact is three-fold: (a) It changes the transconductance of the transconductors from J to aJ. (b) It changes the input coefficients of the third-order Volterra circuit from $\frac{J}{3}$ to $\frac{a^3 J}{3}$. (c) It affects the responses of the first-order Volterra circuit, specifically $v_{1,1}$ and $v_{2,1}$ subsequently $v_{1,1}^3$ and $v_{2,1}^3$, which are part of the inputs of the third-order Volterra circuit.

5.2.3 *Effective Injection Signals*

As the effective injection signals of the first-order Volterra circuit are made of both the external injection signals and the contribution of the third-order Volterra circuit, in what follows we investigate how the nonlinearity factor affects the effective injection signals of the first-order Volterra circuit. Again, we choose the phase of $V_{2,1}$ as the reference phase and let the differential injection currents be

$$i_{in2,1} = I_{inj}\sin(\omega_{inj}t + \phi_{inj}), \tag{5.8a}$$

$$i_{in1,1} = -I_{inj}\sin(\omega_{inj}t + \phi_{inj}). \tag{5.8b}$$

Assume that the frequency of the injection signal $\omega_{inj} = \omega_o + \Delta\omega$ falls into the lock range of the oscillator such that the oscillator will be locked to the injection signal and will oscillate at ω_{inj} once it is locked. The first-order Volterra circuit is linear with its inputs at ω_{inj}. Its outputs therefore only contain frequency components at ω_{inj} and can be written as

$$v_{2,1} = V_{2,1,m}\sin\left(\omega_{inj}t\right), \tag{5.9a}$$

$$v_{1,1} = -V_{1,1,m}\sin\left(\omega_{inj}t\right). \tag{5.9b}$$

Note that the phase of the voltage of node 2 is chosen to be the reference and $v_{1,1}$ and $v_{2,1}$ are out of phase. The delayed version of $v_{2,1}$ and $v_{1,1}$ are given by

$$\hat{v}_{2,1} = V_{2,1,m}\sin\left(\omega_{inj}t - \frac{\pi}{2}\right), \tag{5.10a}$$

$$\hat{v}_{1,1} = -V_{1,1,m}\sin\left(\omega_{inj}t - \frac{\pi}{2}\right). \tag{5.10b}$$

Since

$$\hat{v}_{2,1}^3 = \frac{3}{4}V_{2,1,m}^3\sin\left(\omega_{inj}t - \frac{\pi}{2}\right) - \frac{1}{4}V_{2,1,m}^3\sin\left(3\omega_{inj}t - \frac{3\pi}{2}\right), \tag{5.11a}$$

$$\hat{v}_{1,1}^3 = -\frac{3}{4}V_{1,1,m}^3\sin\left(\omega_{inj}t - \frac{\pi}{2}\right) + \frac{1}{4}V_{1,1,m}^3\sin\left(3\omega_{inj}t - \frac{3\pi}{2}\right), \tag{5.11b}$$

we obtain the inputs of the third-order Volterra circuit given by (5.7)

$$i_{in1,3} = \frac{aJ}{4}V_{2,1,m}^3\sin\left(\omega_{inj}t - \frac{\pi}{2}\right) - \frac{a^3J}{12}V_{2,1,m}^3\sin\left(3\omega_{inj}t - \frac{3\pi}{2}\right), \tag{5.12a}$$

$$i_{in2,3} = -\frac{aJ}{4}V_{1,1,m}^3\sin\left(\omega_{inj}t - \frac{\pi}{2}\right) + \frac{a^3J}{12}V_{1,1,m}^3\sin\left(3\omega_{inj}t - \frac{3\pi}{2}\right). \tag{5.12b}$$

The effective injection signal of the first-order Volterra circuit is obtained by summing up (5.8), the external injection signals, and the components in (5.12) that are at ω_{inj}, which is the contribution of the third-order Volterra circuit to the injection signals of the first-order Volterra circuit

$$
\begin{aligned}
i_{in1,1,eff} &= -I_{inj}\sin\left(\omega_{inj}t+\phi_{inj}\right)+\frac{a^3 J}{4}V_{2,1,m}^3\sin\left(\omega_{inj}t-\frac{\pi}{2}\right)\\
&= -I_{inj}\sin\left(\omega_{inj}t+\phi_{inj}\right)-\frac{a^3 J}{4}V_{2,1,m}^3\cos\left(\omega_{inj}t\right),
\end{aligned}
\tag{5.13a}
$$

$$
\begin{aligned}
i_{in2,1,eff} &= I_{inj}\sin\left(\omega_{inj}t+\phi_{inj}\right)-\frac{a^3 J}{4}V_{1,1,m}^3\sin\left(\omega_{inj}t-\frac{\pi}{2}\right)\\
&= I_{inj}\sin\left(\omega_{inj}t+\phi_{inj}\right)+\frac{a^3 J}{4}V_{1,1,m}^3\cos\left(\omega_{inj}t\right).
\end{aligned}
\tag{5.13b}
$$

It is evident in (5.13) that the effective injection signals of the first-order Volterra circuit are also differential and have the same strength. In order to maximize the effective injection signals of the first-order Volterra circuit, the components of the effective injection signals contributed by the third-order Volterra circuit should be in phase with the external injection signals. To achieve this, a natural choice of the injection phase of the external injection signal at node 2 is

$$
\phi_{inj}^* = \frac{\pi}{2}. \tag{5.14}
$$

The maximum effective injection signals at the optimal injection phase $\phi_{inj}^* = \frac{\pi}{2}$ are obtained from (5.13)

$$
i_{in1,1,eff,max} = -\left(I_{inj}+\frac{a^3 J}{4}V_{2,1,m}^3\right)\cos\left(\omega_{inj}t\right), \tag{5.15a}
$$

$$
i_{in2,1,eff,max} = \left(I_{inj}+\frac{a^3 J}{4}V_{1,1,m}^3\right)\cos\left(\omega_{inj}t\right). \tag{5.15b}
$$

It is evident in (5.15) that increasing the degree of the nonlinearity of the oscillator increases the contribution of the third-order Volterra circuit to the effective injection signals of the first-order Volterra circuit. This in turn increases the lock range of the oscillator.

5.2.4 Lock Range

To derive the lock range of the dual-comparator relaxation oscillator, we modify the governing equations of the first-order Volterra circuit derived earlier with the external injection signals replaced with the effective injection signals obtained in the preceding section

$$sC_1V_{1,1} - aJV_{2,1}e^{-j\frac{\pi}{2}} + I_{in1,1,eff} = 0, \tag{5.16}$$

$$sC_2V_{2,1} - aJV_{1,1}e^{-j\frac{\pi}{2}} + I_{in2,1,eff} = 0. \tag{5.17}$$

Solving for $V_{1,1}$ and $V_{2,1}$ gives

$$V_{1,1} = -\frac{sC_2I_{in1,1,eff} + aJe^{-j\frac{\pi}{2}}I_{in2,1,eff}}{s^2C_1C_2 + (aJ)^2}. \tag{5.18a}$$

$$V_{2,1} = -\frac{sC_1I_{in2,1,eff} + aJe^{-j\frac{\pi}{2}}I_{in1,1,eff}}{s^2C_1C_2 + (aJ)^2}. \tag{5.18b}$$

Define

$$H_{o1} = \frac{aJ}{sC_1}, \tag{5.19a}$$

$$H_{o2} = \frac{aJ}{sC_2}, \tag{5.19b}$$

and

$$H_o = H_{o1}H_{o2}. \tag{5.20}$$

When the injection signals are absent, the oscillator will oscillate at ω_o. Barkhausen criteria must therefore be satisfied at ω_o, i.e.,

$$1 + H_{o1}(\omega_o)H_{o2}(\omega_o) = 0. \tag{5.21}$$

Since $C_1 = C_2 = C$, we have $H_{o1}(\omega_o) = H_{o2}(\omega_o)$. As a result,

$$H_{o1}(\omega_o), H_{o2}(\omega_o) = -j = e^{-j\frac{\pi}{2}}. \tag{5.22}$$

It is interesting to note that (5.22) can also be obtained from the definition of H_{o1} and H_{o2} given earlier, and the fact $H_{o1}H_{o2} = -1$. Utilizing these results, we can write (5.16) as

$$V_{1,1} = -\frac{a}{J}\frac{H_o}{1+H_o}\left(\frac{1}{H_{o2}}I_{in1,1,eff} + e^{-j\frac{\pi}{2}}I_{in2,1,eff}\right), \tag{5.23a}$$

$$V_{2,1} = -\frac{a}{J}\frac{H_o}{1+H_o}\left(\frac{1}{H_{o1}}I_{in2,1,eff} + e^{-j\frac{\pi}{2}}I_{in1,1,eff}\right). \tag{5.23b}$$

Since $\omega_{inj} = \omega_o + \Delta\omega$ with $\Delta\omega \ll \omega_o$, approximating $H_o(\omega_o + \Delta\omega)$ with its first-order Taylor series expansion

$$H_o(\omega_o+\Delta\omega)\approx H_o(\omega_o)+\left(\frac{dH_o}{d\omega}\right)_{\omega_o}\Delta\omega, \tag{5.24}$$

and utilizing

$$H_o(\omega_o+\Delta\omega)\approx H_o(\omega_o)=-1, \tag{5.25}$$

we arrive from (5.23)

$$V_{1,1}\approx\frac{a}{J}\frac{1}{\left(\frac{dH_o}{d\omega}\right)_{\omega_o}\Delta\omega}\left(\frac{1}{H_{o2}}I_{in1,1,eff}+e^{-j\frac{\pi}{2}}I_{in2,1,eff}\right), \tag{5.26a}$$

$$V_{2,1}\approx\frac{a}{J}\frac{1}{\left(\frac{dH_o}{d\omega}\right)_{\omega_o}\Delta\omega}\left(\frac{1}{H_{o1}}I_{in1,1,eff}+e^{-j\frac{\pi}{2}}I_{in2,1,eff}\right). \tag{5.26b}$$

It follows from (5.26)

$$V_{1,1}^*\approx\frac{a}{J}\frac{1}{\left(\frac{dH_o^*}{d\omega}\right)_{\omega_o}\Delta\omega}\left(\frac{1}{H_{o2}^*}I_{in1,1,eff}^*+e^{j\frac{\pi}{2}}I_{in2,1,eff}^*\right), \tag{5.27a}$$

$$V_{2,1}^*\approx\frac{a}{J}\frac{1}{\left(\frac{dH_o^*}{d\omega}\right)_{\omega_o}\Delta\omega}\left(\frac{1}{H_{o1}^*}I_{in1,1,eff}^*+e^{j\frac{\pi}{2}}I_{in2,1,eff}^*\right). \tag{5.27b}$$

We therefore have

$$\begin{aligned}|V_{1,1}|^2\approx\left(\frac{a}{J}\right)^2&\frac{1}{\left(\frac{dH_o}{d\omega}\frac{dH_o^*}{d\omega}\right)_{\omega_o}(\Delta\omega)^2}\\&\times\left(|I_{in2,1,eff}|^2+\frac{|I_{in1,1,eff}|^2}{|H_{o2}|^2}+\frac{e^{j\frac{\pi}{2}}}{H_{o2}}I_{in1,1,eff}I_{in2,1,eff}^*\right.\\&\left.+\frac{e^{-j\frac{\pi}{2}}}{H_{o2}^*}I_{in1,1,eff}^*I_{in2,1,eff}\right).\end{aligned} \tag{5.28}$$

Making use of the quality factor introduced earlier in (3.32), which is repeated here for convenience

$$\left(\frac{dH_o}{d\omega}\frac{dH_o^*}{d\omega}\right)_{\omega_o}=\left[\frac{2Q(\omega_o)}{\omega_o}\right]^2, \tag{5.29}$$

we have from (5.28)

$$|V_{1,1}|^2 \approx \left(\frac{a}{J}\right)^2 \left[\frac{\omega_o}{2Q(\omega_o)}\right]^2 \frac{1}{(\Delta\omega)^2} \times \left(|I_{in2,1,eff}|^2 + \frac{|I_{in1,1,eff}|^2}{|H_{o2}|^2} + \frac{e^{j\frac{\pi}{2}}}{H_{o2}} I_{in1,1,eff} I^*_{in2,1,eff} + \frac{e^{-j\frac{\pi}{2}}}{H^*_{o2}} I^*_{in1,1,eff} I_{in2,1,eff}\right). \tag{5.30}$$

Solve for $\Delta\omega$ from (5.30)

$$|\Delta\omega| \approx \frac{a}{J}\frac{\omega_o}{2Q(\omega_o)}\frac{1}{|V_{1,1}|} \times \sqrt{|I_{in2,1,eff}|^2 + \frac{|I_{in1,1,eff}|^2}{|H_{o2}|^2} + \frac{e^{j\frac{\pi}{2}}}{H_{o2}} I_{in1,1,eff} I^*_{in2,1,eff} + \frac{e^{-j\frac{\pi}{2}}}{H^*_{o2}} I^*_{in1,1,eff} I_{in2,1,eff}}. \tag{5.31}$$

Making use of

$$H_{o1}(\omega_o + \Delta\omega) \approx H_{o1}(\omega_o) = e^{-j\frac{\pi}{2}}, \tag{5.32a}$$

$$H_{o2}(\omega_o + \Delta\omega) \approx H_{o2}(\omega_o) = e^{-j\frac{\pi}{2}}, \tag{5.32b}$$

which was derived earlier in (5.22) and noting that at the optimal injection phase $\phi^*_{inj} = \frac{\pi}{2}$, the effective injection signals are maximized and their phase difference is 180°, i.e.,

$$I_{in1,1,eff} = -I_{eff1}, \tag{5.33a}$$

$$I_{in2,1,eff} = I_{eff1}, \tag{5.33b}$$

we have

$$\begin{aligned} |\Delta\omega| &\approx \frac{a}{J}\frac{\omega_o}{2Q(\omega_o)}\frac{1}{|V_{1,1}|}\sqrt{I^2_{eff1} + I^2_{eff1} + e^{j\pi} I^2_{eff1} + e^{-j\pi} I^2_{eff1}} \\ &\approx \frac{a}{J}\frac{\omega_o}{Q(\omega_o)}\frac{I_{eff1}}{|V_{1,1}|}. \end{aligned} \tag{5.34}$$

For the purpose of comparison, we copy the lock range of the dual-comparator relaxation oscillator with $a = 1$ that was derived in (4.90) here

$$|\Delta\omega| \approx \frac{1}{J}\frac{\omega_o}{Q(\omega_o)}\frac{I_{eff1}}{|V_{1,1}|}. \tag{5.35}$$

In order to have a fair comparison, we assume that $|V_{1,1}|$ is the same in both cases. It is seen from (5.34) and (5.35) that when the degree of the nonlinearity of the dual-comparator relaxation oscillator increases, the lock range of the oscillator increases not only because the lock range is directly proportional to nonlinear factor a, the effective injection signals of the first-order Volterra circuit of the oscillator also increase due to the increased contribution of the high-order Volterra circuits.

5.2.5 Relaxation Oscillators and Ring Oscillators: A Comparison

In this section, we compare the dual-comparator relaxation oscillator and the ring oscillator studied earlier in terms of lock range. Although both relaxation oscillators and ring oscillators fall into the category of nonharmonic oscillators, their intrinsic distinct characteristics specifically the different degrees of nonlinearity give rise to different lock ranges. Our goal of this section is to explore the root causes of their different lock ranges.

The dual-comparator relaxation oscillator can be depicted by the system shown in Fig. 5.5. The frequency components of the output current of the nonlinear transconductors at ω_{inj}, $3\omega_{inj}$, and $5\omega_{inj}$ are attenuated by the first-order low-pass filters formed by the output resistance of the transconductors and the integrating capacitors. Similarly, the ring oscillator with static inverters as delay stages also have a first-order low-pass filter formed by the output resistance of the driving inverter and the load capacitance of the delay stage at the output of each transconductor that models the inverter stage, as shown in Fig. 5.5. Unlike the relaxation oscillator where the nonlinear characteristics of the transconductors are harsh, the characteristics of the nonlinear transconductors of the ring oscillator are rather mild. If the output current of the nonlinear transconductors of both oscillators is depicted using the following hyperbolic tangent function

$$i_{o,relax} = \frac{J}{2}\left[\tanh\left(a_{relax}v\right) + 1\right], \tag{5.36a}$$

$$i_{o,ring} = \frac{J}{2}\left[\tanh\left(a_{ring}v\right) + 1\right], \tag{5.36b}$$

where a_{relax} and a_{ring} are the nonlinearity factor of the relaxation oscillator and ring oscillator, respectively, and v is the input of the transconductors, we will have

$$a_{relax} > a_{ring}. \tag{5.37}$$

In the vicinity of the state transition of the transconductors, tanh(av) can be approximated using its third-order Taylor series expansion. Equation (5.36) becomes

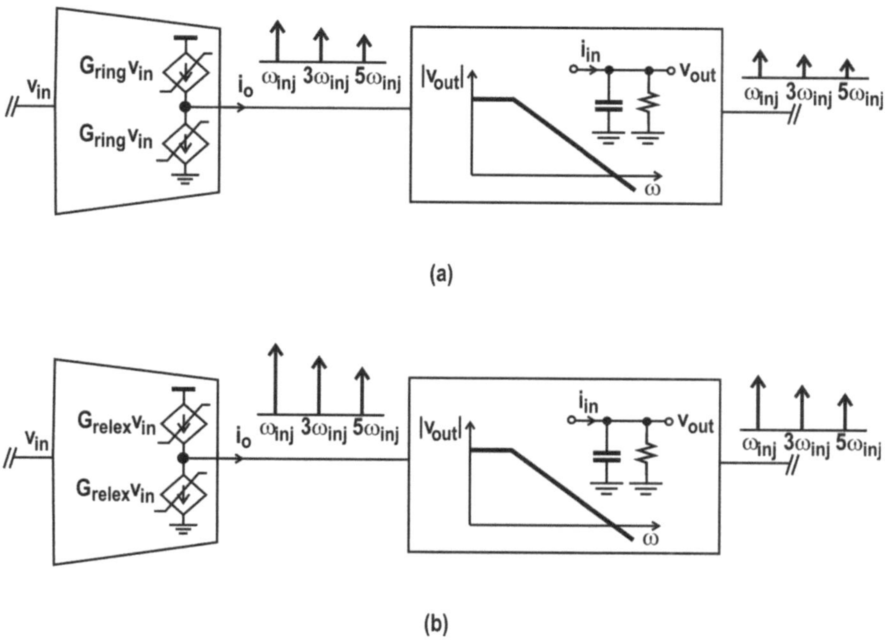

Fig. 5.5 (**a**) Ring oscillators and (**b**) relaxation oscillators have the same low-pass filter at the output of each stage. The harsher nonlinear characteristics of relaxation oscillators give rise to the larger harmonic components of the output current of the transconductors

$$i_{o,relax} \approx \frac{J}{2}\left(a_{relax}v - \frac{a_{relax}^3}{3}v^3 + 1\right), \tag{5.38a}$$

$$i_{o,ring} \approx \frac{J}{2}\left(a_{ring}v - \frac{a_{ring}^3}{3}v^3 + 1\right). \tag{5.38b}$$

Assume that both oscillators have the same nodal voltage profile, i.e., their nodal voltages have the same amplitude and frequency. Further assume that the first-order low-pass filter at the output of the transconductors in both oscillators is the same. Since $a_{relax} > a_{ring}$, the frequency components of $i_{o,relax}$ at ω_{inj}, $3\omega_{inj}$, and $5\omega_{inj}$ are larger as compared with those of $i_{o,ring}$, as illustrated graphically in Fig. 5.5. When $i_{o,relax}$ is converted to a voltage via the low-pass filter, the resultant voltages at ω_{inj}, $3\omega_{inj}$, and $5\omega_{inj}$ will be larger as compared with those generated by $i_{o,ring}$. As a result, the effective injection signals subsequently the lock range of the relaxation oscillator are larger as compared with those of the ring oscillator.

To validate the preceding theoretical findings, consider 2 three-stage ring oscillators shown in Fig. 5.6, one with purely capacitive loads and the other with RC

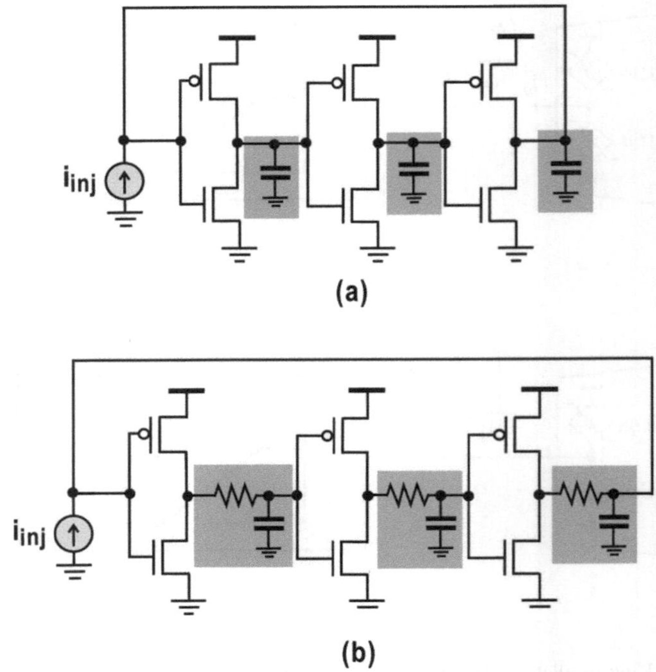

Fig. 5.6 Ring oscillators with (**a**) capacitive loads and (**b**) RC loads

loads [113]. By adjusting the RC time constant, we can control the −3 dB frequency of the low-pass filter so as to adjust the amount of harmonics that contribute to the effective injection signals of the first-order Volterra circuit of the oscillator. To obtain a fair comparison of the lock range, the two oscillators have the same transistor dimensions and oscillation frequency. Also, both oscillators have the same injection ratio, which is the ratio of the injection current to the total current drawn by the oscillators from their power supply. The free running frequency of the both oscillators is set to 253 MHz, obtained by varying the supply voltage of the oscillators. The capacitive load of the oscillator Fig. 5.6a is made of the device capacitances. No explicit capacitor is added. The RC load of the oscillator Fig. 5.6b is explicitly added. The simulated performance of the oscillators is compared in Table 5.1. It is seen that the ring oscillator with purely capacitive loads has smaller normalized harmonics while that with RC loads has larger normalized harmonics. The amplitude of harmonics is normalized with respect to that of the first harmonic. The lock range of the oscillator with RC loads is larger as compared with that with capacitive loads, agreeing well with the theoretical findings presented earlier.

Table 5.1 Comparison of the performance of ring oscillators in Fig. 5.6

Oscillator	Figure 5.6b	Figure 5.6a
Tech.	130 nm 1.2V CMOS	130 nm 1.2 V CMOS
Free-running freq. ω_o	253 MHz	253 MHz
Third norm. harmonic	−15.22 dB	−20.84 dB
Fifth norm. harmonic	−17.37 dB	−23.18 dB
Injection ratio	0.2	0.2
Lock ranges	59 MHz	35 MHz

5.3 Single Unitone Injection versus Dual Unitone Injections

In Chap. 4, we studied the injection-locked dual-comparator relaxation oscillator with dual current injections, specifically, the injection of a pair of differential currents. In many applications only a single-ended injection is available and the conversion from a single-ended signal to a pair of differential signals might be costly. The characteristics of nonharmonic oscillators with only one injection is clearly of great interest. In this section, we investigate the injection-locked dual-comparator relaxation oscillator with only one injection. It can be considered as a special case of the injection-locked dual-comparator relaxation oscillator with dual injections studied in Chap. 4 such that the results obtained for the dual-comparator injection-locked relaxation oscillator with dual injections can be utilized directly.

Consider the dual-comparator relaxation oscillator with current i_{inj} injected into node 2, as shown in Fig. 5.7. The governing equations of the oscillator are obtained from Kirchhoff's Current Law at nodes 1 and 2 while accounting for the relation between the complementary outputs of the SR latch

$$C_1\frac{dv_1}{dt} - 2JB_2 + J = 0, \tag{5.39a}$$

$$C_2\frac{dv_2}{dt} - 2JB_1 + J + i_{inj} = 0, \tag{5.39b}$$

$$B_1 + B_2 = 1. \tag{5.39c}$$

The third constraint in (5.39) can be accounted for by the insertion of two delay blocks in the block diagram, as shown in Fig. 5.8. The exact approach used to analyze the dual-comparator relaxation oscillator with dual injection presented earlier can be used to obtain the effective injection signals of the first-order Volterra circuit of the oscillator. The effective injection signal given by (4.67) for the oscillator with dual injections is repeated here for convenience

$$\begin{aligned} I_{in1,1,eff} &= -I_{inj}\sin\left(\omega_{inj}t + \phi_{inj}\right) + \frac{J}{4}V_{2,1,m}^3 \sin\left(\omega_{inj}t - \frac{\pi}{2}\right) \\ &= -I_{inj}\sin\left(\omega_{inj}t + \phi_{inj}\right) - \frac{J}{4}V_{2,1,m}^3 \cos\left(\omega_{inj}t\right), \end{aligned} \tag{5.40a}$$

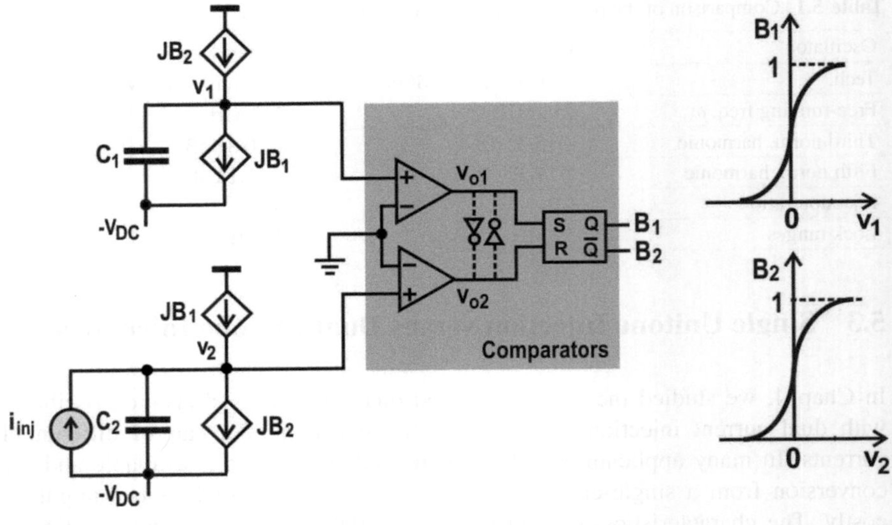

Fig. 5.7 Dual-comparator relaxation oscillator with single injection

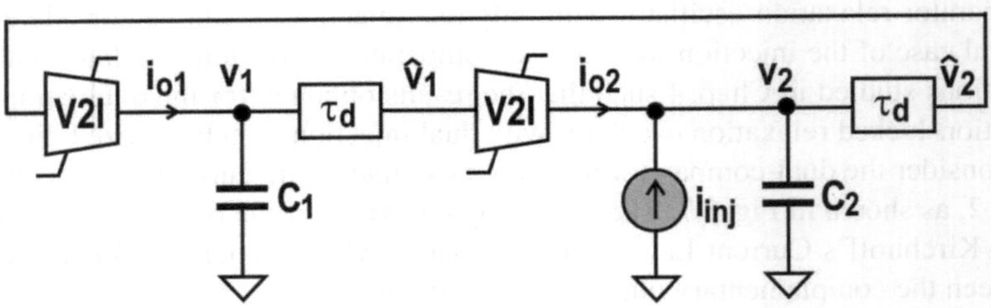

Fig. 5.8 Block diagram of dual-comparator relaxation oscillator with single injection

$$I_{in2,1,eff} = I_{inj}\sin\left(\omega_{inj}t + \phi_{inj}\right) - \frac{J}{4}V_{1,1,m}^3 \sin\left(\omega_{inj}t - \frac{\pi}{2}\right)$$

$$= I_{inj}\sin\left(\omega_{inj}t + \phi_{inj}\right) + \frac{J}{4}V_{1,1,m}^3 \cos\left(\omega_{inj}t\right). \tag{5.40b}$$

Since the external injection is only applied to node 2, Eq. (5.40) is modified to

$$I_{in1,1,eff} = -\frac{J}{4}V_{2,1,m}^3 \cos\left(\omega_{inj}t\right), \tag{5.41a}$$

$$I_{in2,1,eff} = I_{inj}\sin\left(\omega_{inj}t + \phi_{inj}\right) + \frac{J}{4}V_{1,1,m}^3 \cos\left(\omega_{inj}t\right). \tag{5.41b}$$

In order to maximize the effective injection signals of the first-order Volterra circuit, the components contributed by the third-order Volterra circuit should be in phase with the external injection signal. To achieve this, the injection phase of the external injection signal is set to

$$\phi_{inj}^* = \frac{\pi}{2}. \tag{5.42}$$

The maximum effective injection signal at the optimal injection phase is therefore given by

$$I_{in1,1,eff,max} = -\frac{J}{4} V_{2,1,m}^3 \cos\left(\omega_{inj} t\right), \tag{5.43a}$$

$$I_{in2,1,eff,max} = \left(I_{inj} + \frac{J}{4} V_{1,1,m}^3\right) \cos\left(\omega_{inj} t\right). \tag{5.43b}$$

It is seen from (5.43) that even though there is no external injection signal at node 1, the third-order Volterra circuit contributes to the effective injection signals at both nodes 1 and 2. Also, the two effective injection signals have a phase difference of 180°. Following the same approach as that for deriving the lock range of the oscillator with dual injections, one can obtain the lock range of the oscillator with only single injection

$$|\Delta\omega| \approx \frac{1}{J} \frac{\omega_o}{2Q(\omega_o)} \frac{1}{|V_{1,1}|} \times \sqrt{|I_{in2,1,eff}|^2 + \frac{|I_{in1,1,eff}|^2}{|H_{o2}|^2} + \frac{e^{j\frac{\pi}{2}}}{H_{o2}} I_{in1,1,eff} I_{in2,1,eff}^* + \frac{e^{-j\frac{\pi}{2}}}{H_{o2}^*} I_{in1,1,eff}^* I_{in2,1,eff}}. \tag{5.44}$$

If the contribution of the third-order Volterra circuit to the effective injection signals of the first-order Volterra circuit is much smaller as compared with the external injection signal, i.e., $I_{in1,1,eff,max} \ll I_{in2,1,eff,max}$, Eq. (5.44) can be simplified to

$$|\Delta\omega| \approx \frac{1}{2} \frac{1}{J} \frac{\omega_o}{Q(\omega_o)} \frac{|I_{inj,1,eff}|}{|V_{1,1}|}. \tag{5.45}$$

A comparison of (5.45) with the lock range with dual injections given by (4.90) and is repeated in (5.46)

$$|\Delta\omega| \approx \frac{1}{J} \frac{\omega_o}{Q(\omega_o)} \frac{I_{eff1}}{|V_{1,1}|} \tag{5.46}$$

reveals that the lock range of the dual-comparator relaxation oscillator with dual injections is twice that with only one injection, provided that the two injection

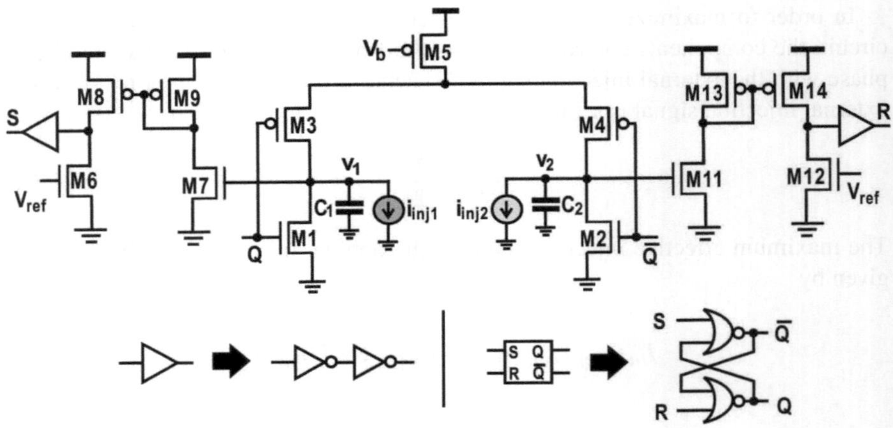

Fig. 5.9 Dual-comparator relaxation oscillator

signals be injected into the oscillator with proper injection phases, specifically, the injection phase of the current injected into node 2 of the oscillator is $\frac{\pi}{2}$ and the current injected into node 1 of the oscillator is out of phase with that injected into node 2.

As an example, consider the dual-comparator relaxation oscillator shown in Fig. 5.9 designed in IBM 130 nm 1.2 V CMOS technology [106]. The oscillator operates in the sub-threshold mode achieved by properly biasing voltages and oscillates at 13 MHz. Both single unitone injection $i_{inj} = I_{inj}\sin(\omega_{inj}t)$ and dual unitone injections $i_{inj1} = I_{inj}\sin(\omega_{inj}t)$ and $i_{inj2} = -I_{inj}\sin(\omega_{inj}t)$ are considered. For dual unitone injections, the two injection currents have the same frequency and amplitude but a 180° phase difference.

Figure 5.10 plots the dependence of the lock range of the oscillator on injection strength, which is the ratio of the injection current to the current of transistor M5. It is seen from the figure that the lock range is directly proportional to injection strength. Also observed is that for the same injection strength, the lock range of the oscillator with dual injections is approximately twice that with a single injection, agreeing well with the theoretical results presented earlier.

5.4 Dual Multitone Injections

In Chap. 4.2, we investigated the injection-locked nonharmonic oscillators with a unitone injection. We showed that the lock range of a nonharmonic oscillator is the same as that of its first-order Volterra circuit. We also showed that the effective injection signals of the first-order Volterra circuit of the nonharmonic oscillator at the frequency of the external injection signal consists of (a) the external injection

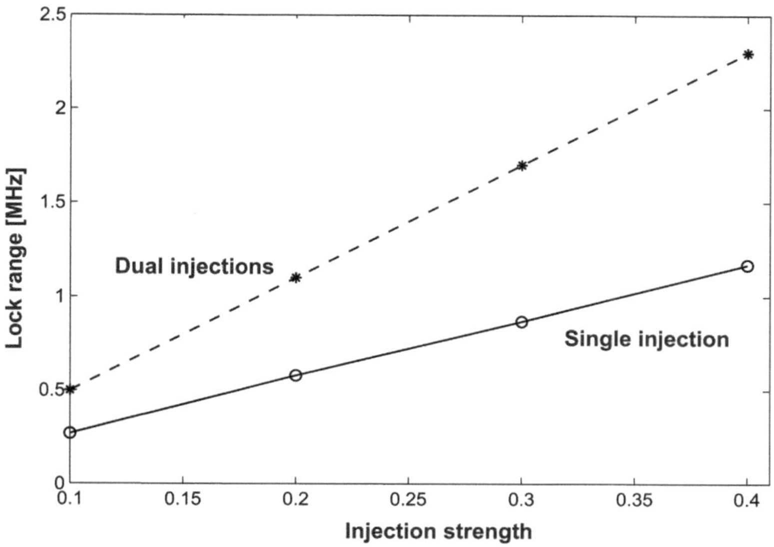

Fig. 5.10 Simulated lock range of dual-comparator relaxation oscillator in Fig. 5.9 with single unitone injection and dual unitone injections. The phase difference between the dual injection signals is 180°

signals and (b) the contribution of the high-order Volterra circuits of the oscillator. We further showed in Sect. 5.2 that the contribution of the high-order Volterra circuits of the oscillator is directly related to the degree of the nonlinearity of the oscillator. The more severe the nonlinearity of the oscillator, the more the high-order Volterra circuits of the oscillator contribute to the effective injection signal of the first-order Volterra circuit of the oscillator and subsequently the larger the lock range of the oscillator.

In applications such as a serial data link where a frequency synthesizer is typically used to generate the system clocks that control the operation of both serialization and pre-emphasis of the transmitter, the frequency of the ring oscillator of the frequency synthesizer is typically in a multi-GHz range while that of the reference clock is only a few MHz as the reference clock is often provided by a crystal oscillator to take the advantage of the superior phase noise of the oscillator. A frequency divider with a large frequency division ratio in the feedback path of the synthesizer is needed to generate a low-frequency signal from the output of the ring oscillator so that it can compare with the reference clock. The high frequency division ratio of the frequency divider requires that frequency division be realized in a multistage fashion with the first stage known as the pre-scalar typically realized using an injection-locked frequency divider. The injection signal of the pre-scalar of the frequency synthesizer is the output of the ring oscillator of the synthesizer that contains an infinite number of frequency components whose frequency is harmonically related. In this section, we investigate whether the direct output of the ring oscillator of the synthesizer, which is a multitone signal, or the first harmonic of the output of the ring oscillator of the synthesizer, which is a unitone signal, should be used as the injection signal of the pre-scalar. We want to know which one gives a larger lock range.

5.4.1 Multitone Signals

Consider an injection current i_{inj} whose time-domain waveform and frequency-domain spectrum are shown in Fig. 5.11. The Fourier series expansion of i_{inj} is given by

$$i_{inj}(t) = \frac{I_m}{2} + \frac{4I_m}{\pi} \sum_{k=1,3,5,...}^{\infty} \frac{1}{k} \sin(k\omega_{inj}t), \tag{5.47}$$

where $\omega_{inj} = 2\pi/T_{inj}$ and T_{inj} is the period of the injection signal. It is seen that the injection signal only contains odd harmonics. The spectrum of the output of the oscillator consists of a train of impulses at frequencies $\omega_{inj}, 3\omega_{inj}, 5\omega_{inj}, \ldots$, whose amplitude drops with the increase in the order of the harmonic in a hyperbolic manner, as shown in Fig. 5.11. Since the amplitude of the harmonics drops with the order of the harmonics, to simplify analysis, we only consider the first three terms of the Fourier series expansion of i_{inj}

$$i_{inj} \approx I_{inj,1} \sin(\omega_{inj}t) + I_{inj,3} \sin(3\omega_{inj}t) + I_{inj,5} \sin(5\omega_{inj}t), \tag{5.48}$$

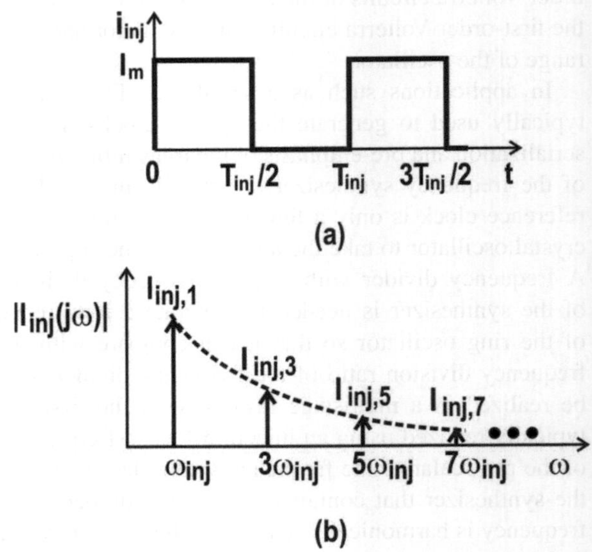

Fig. 5.11 (**a**) Waveform and (**b**) spectrum of a square-wave injection signal with zero injection phase

where

$$I_{inj,1} = \frac{4I_m}{\pi}, \tag{5.49a}$$

$$I_{inj,3} = \frac{4I_m}{3\pi}, \tag{5.49b}$$

$$I_{inj,5} = \frac{4I_m}{5\pi}. \tag{5.49c}$$

Note we have dropped the DC component in (5.47) as it has no impact on the frequency of the oscillator under injection.

If the injection current has a nonzero injection phase ϕ_{inj}, as shown in Fig. 5.12, since

$$\phi_{inj} = \omega_{inj} t_{inj}, \tag{5.50}$$

we have

$$t_{inj} = \frac{\phi_{inj}}{\omega_{inj}}. \tag{5.51}$$

To show the impact of the injection phase ϕ_{inj}, we rewrite (5.47) in its exponential form with the aid of Euler's formula

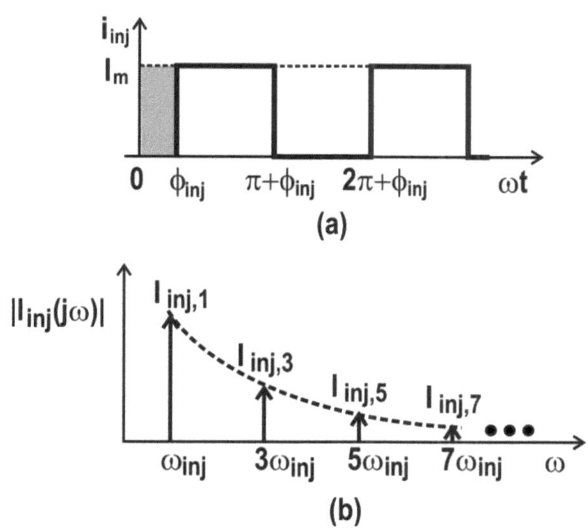

Fig. 5.12 (**a**) Waveform and (**b**) spectrum of a square-wave injection signal with injection phase ϕ_{inj}

$$i_{inj}(t) = \frac{I_m}{2} + \frac{2I_m}{j\pi} \sum_{k=1,3,5,\ldots}^{\infty} \frac{1}{k} \left(e^{jk\omega_{inj}t} - e^{-jk\omega_{inj}t} \right). \tag{5.52}$$

If the injection current has a nonzero injection phase ϕ_{inj}, as shown in Fig. 5.12, Eq. (5.52) needs to be modified to

$$i_{inj}(t) = \frac{I_m}{2} + \frac{2I_m}{j\pi} \sum_{k=1,3,5,\ldots}^{\infty} \frac{1}{k} \left[e^{jk\omega_{inj}(t-t_{inj})} - e^{-jk\omega_{inj}(t-t_{inj})} \right]. \tag{5.53}$$

Making use of (5.51), we arrive at

$$\begin{aligned} i_{inj}(t) &= \frac{I_m}{2} + \frac{2I_m}{j\pi} \sum_{k=1,3,5,\ldots}^{\infty} \frac{1}{k} \left(e^{jk\omega_{inj}t} - e^{-jk\omega_{inj}t} \right) e^{-jk\phi_{inj}} \\ &= \frac{I_m}{2} + \frac{4I_m}{\pi} \sum_{k=1,3,5,\ldots}^{\infty} \frac{1}{k} \sin\left(k\omega_{inj}t - k\phi_{inj} \right). \end{aligned} \tag{5.54}$$

Equation (5.54) reveals that if a square-wave injection signal has an injection phase ϕ_{inj}, the corresponding injection phase of the first, third, and fifth harmonics of the injection signal will be ϕ_{inj}, $3\phi_{inj}$, and $5\phi_{inj}$, respectively. Also the injection phase does not affect the amplitude of the harmonics of the injection signal.

5.4.2 Volterra Circuits

Let us now consider the dual-comparator relaxation oscillator studied earlier, this time with multitone injection. The external differential injection currents are 50% duty cycle square waves. The governing equations of the oscillator are the same as those with unitone injection given in (4.50) and are repeated here for convenience

$$C_1 \frac{dv_1}{dt} - JB_2 + J + i_{inj1} = 0, \tag{5.55a}$$

$$C_2 \frac{dv_2}{dt} - JB_1 + J + i_{inj2} = 0, \tag{5.55b}$$

$$B_1 + B_2 = 1. \tag{5.55c}$$

We follow the same approach used for analyzing the dual-comparator relaxation oscillator with unitone injection to analyze the oscillator with multitone injections given by

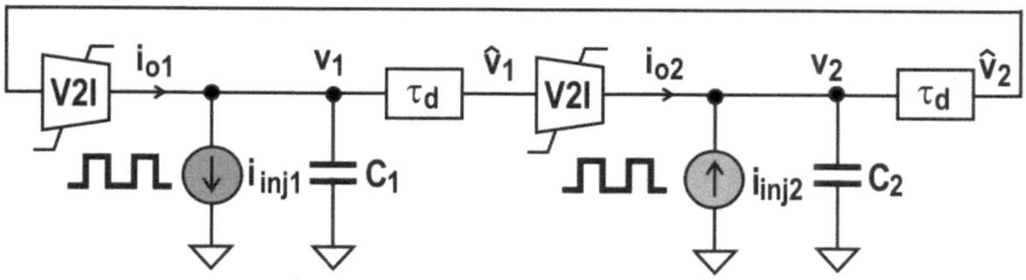

Fig. 5.13 Dual-comparator relaxation oscillator with dual multitone injections

$$i_{inj} \approx I_{inj,1}\sin(\omega_{inj}t+\phi_{inj}) + I_{inj,3}\sin(3\omega_{inj}t+3\phi_{inj}) + I_{inj,5}\sin(5\omega_{inj}t+5\phi_{inj}). \tag{5.56}$$

It was shown in (5.54) that ϕ_{inj} is the injection phase of the square-wave injection current, which is the same as that of the first harmonic of the injection signal. The injection phase of the third harmonic and that of the fifth harmonic are $3\phi_{inj}$ and $5\phi_{inj}$, respectively. The third constraint in (5.55) is accounted for by the addition of two delay blocks shown in Fig. 5.13.

Let the injection signal be perturbed slightly such that the amplitude of the injection signal is changed from I_m to ϵI_m while the frequency of the injection signal remains unchanged. The injection signal in this case becomes

$$i_{inj} \approx \epsilon I_{inj,1}\sin(\omega_{inj}t+\phi_{inj}) + \epsilon I_{inj,3}\sin(3\omega_{inj}t+3\phi_{inj}) + \epsilon I_{inj,5}\sin(5\omega_{inj}t+5\phi_{inj}). \tag{5.57}$$

As before, the relaxation oscillator is represented by its first-, second-, and third-order Volterra circuits with their governing equations given by

$$C_1\frac{dv_{1,k}}{dt} - J\hat{v}_{2,k} + i_{in1,k} = 0, \tag{5.58a}$$

$$C_2\frac{dv_{2,k}}{dt} - J\hat{v}_{1,k} + i_{in2,k} = 0, \tag{5.58b}$$

where $k = 1, 2, 3$. The inputs of the first-, second-, and third-order Volterra circuits are given by

$$i_{in1,1} = -I_{inj,1}\sin(\omega_{inj}t+\phi_{inj}) - I_{inj,3}\sin(3\omega_{inj}t+3\phi_{inj}) - I_{inj,5}\sin(5\omega_{inj}t+5\phi_{inj}), \tag{5.59a}$$

$$i_{in2,1} = I_{inj,1}\sin(\omega_{inj}t+\phi_{inj}) + I_{inj,3}\sin(3\omega_{inj}t+3\phi_{inj}) + I_{inj,5}\sin(5\omega_{inj}t+5\phi_{inj}), \tag{5.59b}$$

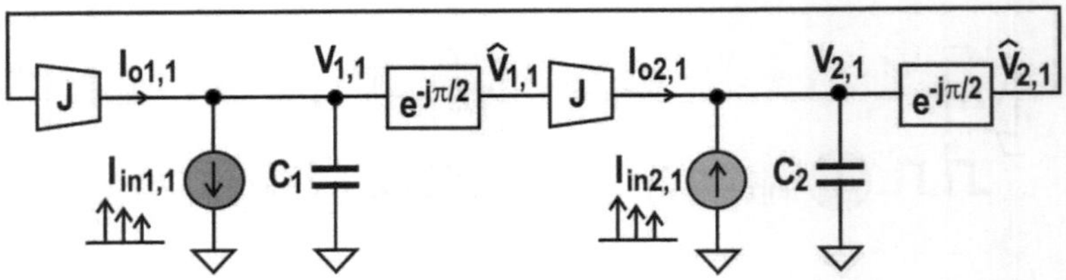

Fig. 5.14 First-order Volterra circuit of dual-comparator relaxation oscillator with dual multitone injections. The contributions of the higher-order Volterra circuits are not accounted for. Should the impact of high-order Volterra circuits be accounted for, the injection signals would consist of both the external multitone injection signals and the contribution of the high-order Volterra circuits

$$i_{in1,2} = 0, \tag{5.59c}$$

$$i_{in2,2} = 0, \tag{5.59d}$$

$$i_{in1,3} = \frac{J}{3}\hat{v}_{2,1}^3, \tag{5.59e}$$

$$i_{in2,3} = \frac{J}{3}\hat{v}_{1,1}^3. \tag{5.59f}$$

The opposite signs of $i_{in1,1}$ and $i_{in2,1}$ indicate that the currents are differentially injected into the oscillator. Figure 5.14 shows the block diagram of the first-order Volterra circuit of the dual-comparator relaxation oscillator with multitone injection in the phasor domain. Note that the injection signals are multitoned.

5.4.3 *Effective Injection Signals*

The inputs of the first-order Volterra circuits are the external injection signals that contain frequency components at ω_{inj}, $3\omega_{inj}$, and $5\omega_{inj}$. The response of the first-order Volterra circuit in the lock state only contains frequency components at ω_{inj}, $3\omega_{inj}$, and $5\omega_{inj}$, and can be written as

$$v_{1,1} \approx -V_{1,1,1,m}\sin(\omega_{inj}t) - V_{1,1,3,m}\sin(3\omega_{inj}t) - V_{1,1,5,m}\sin(5\omega_{inj}t), \tag{5.60a}$$

$$v_{2,1} \approx V_{2,1,1,m}\sin(\omega_{inj}t) + V_{2,1,3,m}\sin(3\omega_{inj}t) + V_{2,1,5,m}\sin(5\omega_{inj}t), \tag{5.60b}$$

where $V_{n,1,k,m}$ denotes the amplitude of the kth harmonic of the voltage of node n of the first-order Volterra circuit. Subscript m indicates the maximum amplitude. Note that the phase of the voltage of node 2 is chosen to be the reference and the voltage of node 1 and that of node 2 have a phase difference of 180°. The delayed version of $v_{2,1}$ and $v_{1,1}$ is given by

$$\hat{v}_{1,1} \approx -V_{1,1,1,m} \sin\left(\omega_{inj}t - \frac{\pi}{2}\right) - V_{1,1,3,m} \sin\left(3\omega_{inj}t - \frac{3\pi}{2}\right)$$
$$- V_{1,1,5,m} \sin\left(5\omega_{inj}t - \frac{5\pi}{2}\right), \tag{5.61a}$$

$$\hat{v}_{2,1} \approx V_{2,1,1,m} \sin\left(\omega_{inj}t - \frac{\pi}{2}\right) + V_{2,1,3,m} \sin\left(3\omega_{inj}t - \frac{3\pi}{2}\right)$$
$$+ V_{2,1,5,m} \sin\left(5\omega_{inj}t - \frac{5\pi}{2}\right). \tag{5.61b}$$

For the second-order Volterra circuit, because the inputs of the second-order Volterra are zero, the responses of the second-order Volterra circuit are zero as well. To obtain the inputs of the third-order Volterra circuit, since

$$\begin{aligned}
v_{1,1}^3 = &-V_{1,1,1,,m}^3 \sin^3\left(\omega_{inj}t - \frac{\pi}{2}\right)\\
&- V_{1,1,3,m}^3 \sin^3\left(3\omega_{inj}t - \frac{3\pi}{2}\right)\\
&- V_{1,1,5,m}^3 \sin^3\left(5\omega_{inj}t - \frac{5\pi}{2}\right)\\
&- 3V_{1,1,1}^2 V_{1,1,3,m} \sin^2\left(\omega_{inj}t - \frac{\pi}{2}\right)\sin\left(3\omega_{inj}t - \frac{3\pi}{2}\right)\\
&- 3V_{1,1,1,m}^2 V_{1,1,5} \sin^2\left(\omega_{inj}t - \frac{\pi}{2}\right)\sin\left(5\omega_{inj}t - \frac{5\pi}{2}\right)\\
&- 3V_{1,1,3,m}^2 V_{1,1,1} \sin^2\left(3\omega_{inj}t - \frac{3\pi}{2}\right)\sin\left(\omega_{inj}t - \frac{\pi}{2}\right)\\
&- 3V_{1,1,3}^2 V_{1,1,5,m} \sin^2\left(3\omega_{inj}t - \frac{3\pi}{2}\right)\sin\left(5\omega_{inj}t - \frac{5\pi}{2}\right)\\
&- 3V_{1,1,5,m}^2 V_{1,1,1} \sin^2\left(5\omega_{inj}t - \frac{5\pi}{2}\right)\sin\left(\omega_{inj}t - \frac{\pi}{2}\right)\\
&- 3V_{1,1,5}^2 V_{1,1,3,m} \sin^2\left(5\omega_{inj}t - \frac{\pi}{2}\right)\sin\left(3\omega_{inj}t - \frac{3\pi}{2}\right),
\end{aligned} \tag{5.62a}$$

$$\begin{aligned}
v_{2,1}^3 = &V_{2,1,1,,m}^3 \sin^3\left(\omega_{inj}t - \frac{\pi}{2}\right)\\
&+ V_{2,1,3,m}^3 \sin^3\left(3\omega_{inj}t - \frac{3\pi}{2}\right)
\end{aligned}$$

$$+ V_{2,1,5,m}^3 \sin^3\left(5\omega_{inj}t - \frac{5\pi}{2}\right)$$

$$+3V_{2,1,1}^2 V_{2,1,3,m}\sin^2\left(\omega_{inj}t - \frac{\pi}{2}\right)\sin\left(3\omega_{inj}t - \frac{3\pi}{2}\right)$$

$$+3V_{2,1,1,m}^2 V_{2,1,5}\sin^2\left(\omega_{inj}t - \frac{\pi}{2}\right)\sin\left(5\omega_{inj}t - \frac{5\pi}{2}\right)$$

$$+3V_{2,1,3,m}^2 V_{2,1,1}\sin^2\left(3\omega_{inj}t - \frac{3\pi}{2}\right)\sin\left(\omega_{inj}t - \frac{\pi}{2}\right)$$

$$+3V_{2,1,3}^2 V_{2,1,5,m}\sin^2\left(3\omega_{inj}t - \frac{3\pi}{2}\right)\sin\left(5\omega_{inj}t - \frac{5\pi}{2}\right)$$

$$+3V_{2,1,5,m}^2 V_{2,1,1}\sin^2\left(5\omega_{inj}t - \frac{5\pi}{2}\right)\sin\left(\omega_{inj}t - \frac{\pi}{2}\right)$$

$$+3V_{2,1,5}^2 V_{2,1,3,m}\sin^2\left(5\omega_{inj}t - \frac{\pi}{2}\right)\sin\left(3\omega_{inj}t - \frac{3\pi}{2}\right), \quad (5.62b)$$

making use of the following identities of trigonometric functions

$$\sin^2\alpha = \frac{1}{2}(1 - \cos 2\alpha), \quad (5.63a)$$

$$\sin^3\alpha = \frac{1}{4}(3\sin\alpha - \sin 3\alpha), \quad (5.63b)$$

$$\sin\alpha\sin\beta = \frac{1}{2}[-\cos(\alpha+\beta) + \cos(\alpha-\beta)], \quad (5.63c)$$

$$\cos\alpha\cos\beta = \frac{1}{2}[\sin(\alpha+\beta) + \cos(\alpha-\beta)], \quad (5.63d)$$

$$\sin\alpha\cos\beta = \frac{1}{2}[\sin(\alpha+\beta) + \sin(\alpha-\beta)], \quad (5.63e)$$

$$\cos\alpha\sin\beta = \frac{1}{2}[\sin(\alpha+\beta) - \sin(\alpha-\beta)], \quad (5.63f)$$

we obtain the frequency components of $v_{1,1}^3$ and $v_{2,1}^3$ that are at ω_{inj}. They are tabulated in Table 5.2.

The effective injection signal of the first-order Volterra circuit of the oscillator at ω_{inj} is obtained by summing up the external injection signal at ω_{inj} and those contributed by the third-order Volterra circuit at ω_{inj}

$$i_{in1,1,eff} = -I_{inj,1}\sin(\omega_{inj}t + \phi_{inj}) \quad (5.64a)$$

$$+\frac{J}{8}\left(V_{2,1,1,m}^3 + V_{2,1,1,m}^2 V_{2,1,3,m} + V_{2,1,3,m}^2 V_{2,1,5,m}\right)\sin\left(\omega_{inj}t - \frac{\pi}{2}\right),$$

Table 5.2 Frequency components at ω_{inj} that are generated by $v_{1,1}^3$ and $v_{2,1}^3$

	Terms of $V_{1,1}^3$ or $V_{2,1}^3$	Frequency components at ω_{inj}
$V_{1,1}^3$	$-V_{1,1,1,m}^3 \sin^3\left(\omega_{inj}t - \frac{\pi}{2}\right)$	$-\frac{3}{4}V_{1,1,1,m}^3 \sin\left(\omega_{inj}t - \frac{\pi}{2}\right)$
	$-3V_{1,1,1,m}^2 V_{1,1,3,m} \sin^2\left(\omega_{inj}t - \frac{\pi}{2}\right)\sin\left(3\omega_{inj}t - \frac{3\pi}{2}\right)$	$-\frac{3}{4}V_{1,1,1,m}^2 V_{1,1,3,m} \sin\left(\omega_{inj}t - \frac{\pi}{2}\right)$
	$-3V_{1,1,3,m}^2 V_{1,1,5} \sin^2\left(3\omega_{inj}t - \frac{3\pi}{2}\right)\sin\left(5\omega_{inj}t - \frac{5\pi}{2}\right)$	$-\frac{3}{4}V_{1,1,3,m}^2 V_{1,1,5} \sin\left(\omega_{inj}t - \frac{\pi}{2}\right)$
$V_{2,1}^3$	$V_{2,1,1,m}^3 \sin^3\left(\omega_{inj}t - \frac{\pi}{2}\right)$	$\frac{3}{4}V_{2,1,1,m}^3 \sin\left(\omega_{inj}t - \frac{\pi}{2}\right)$
	$3V_{2,1,1,m}^2 V_{2,1,3,m} \sin^2\left(\omega_{inj}t - \frac{\pi}{2}\right)\sin\left(3\omega_{inj}t - \frac{3\pi}{2}\right)$	$\frac{3}{4}V_{2,1,1,m}^2 V_{2,1,3,m} \sin\left(\omega_{inj}t - \frac{\pi}{2}\right)$
	$3V_{2,1,3,m}^2 V_{2,1,5,m} \sin^2\left(3\omega_{inj}t - \frac{3\pi}{2}\right)\sin\left(5\omega_{inj}t - \frac{5\pi}{2}\right)$	$\frac{3}{4}V_{2,1,3,m}^2 V_{2,1,5,m} \sin\left(\omega_{inj}t - \frac{\pi}{2}\right)$

$$i_{in2,1,eff} = I_{inj,1}\sin(\omega_{inj}t + \phi_{inj}) - \frac{J}{8}\left(V_{1,1,1,m}^3 + V_{1,1,1,m}^2 V_{1,1,3,m} + V_{1,1,3,m}^2 V_{1,1,5,m}\right)\sin\left(\omega_{inj}t - \frac{\pi}{2}\right). \tag{5.64b}$$

The effective injection signals of the first-order Volterra circuit is a function of injection phase ϕ_{inj} of the external injection signals. They are maximized when

$$\phi_{inj}^* = \frac{\pi}{2}. \tag{5.65}$$

This is the case where the external injection currents are injected at the threshold-crossings of the output of the oscillator and the third-order Volterra circuit contributes the most to the effective injection signals of the first-Volterra circuit. The effective injection signals in this case are given by

$$i_{in1,1,eff,max} = -\left[I_{inj,1} + \frac{J}{4}\left(V_{2,1,1,m}^3 + V_{2,1,1,m}^2 V_{2,1,3,m} + V_{2,1,3,m}^2 V_{2,1,5,m}\right)\right] \times \cos(\omega_{inj}t), \tag{5.66a}$$

$$i_{in2,1,eff,max} = \left[I_{inj,1} + \frac{J}{4}\left(V_{1,1,1,m}^3 + V_{1,1,1,m}^2 V_{1,1,3,m} + V_{1,1,3,m}^2 V_{1,1,5,m}\right)\right] \times \cos(\omega_{inj}t). \tag{5.66b}$$

To compare with the effective injection signals of the same oscillator with dual unitone injections, we rewrite (5.15) with $a = 1$ here

$$I_{in1,1,eff,max} = -\left(I_{inj} + \frac{J}{4}V_{2,1,m}^3\right)\cos\left(\omega_{inj}t\right), \tag{5.67a}$$

$$I_{in2,1,eff,max} = \left(I_{inj} + \frac{J}{4}V_{1,1,m}^3\right)\cos\left(\omega_{inj}t\right). \tag{5.67b}$$

It is seen from (5.66) and (5.67) that the first, third, and fifth harmonics of the response of the first-order Volterra circuit due to the first, third, and fifth harmonics of the injection signals contribute to the effective injection signals of the first-order Volterra circuit. It is the terms associated with the third and fifth harmonics of the response of the first-order Volterra circuit that contribute to the effective injection signals. As a result, the effective injection signals of the first-order Volterra circuit of the oscillator with multitone injections are larger as compared with those with unitone injections.

5.4.4 Lock Range

Once we have the effective injection signals of the first-order Volterra circuit at ω_{inj}, we are now in the position to obtain the lock range of the first-order Volterra circuit of the oscillator. The external differential currents are injected into the oscillator with the optimal injection phase $\phi^*_{inj} = \frac{\pi}{2}$ and the maximum effective injection signals of the first-order Volterra are given by (5.66). The block diagram of the first-order Volterra circuit with the effective injection signals at ω_{inj} is shown in Fig. 5.15. Write KCL at nodes 1 and 2

$$sC_1V_{1,1} - JV_{2,1}e^{-j\frac{\pi}{2}} + I_{in1,1,eff} = 0, \tag{5.68}$$

$$sC_2V_{2,1} - JV_{1,1}e^{-j\frac{\pi}{2}} + I_{in2,1,eff} = 0. \tag{5.69}$$

Solving for $V_{1,1}$ and $V_{2,1}$ gives

$$V_{1,1} = -\frac{sC_2I_{in1,1,eff} + Je^{-j\frac{\pi}{2}}I_{in2,1,eff}}{s^2C_1C_2 + J^2}, \tag{5.70a}$$

$$V_{2,1} = -\frac{sC_1I_{in2,1,eff} + Je^{-j\frac{\pi}{2}}I_{in1,1,eff}}{s^2C_1C_2 + J^2}. \tag{5.70b}$$

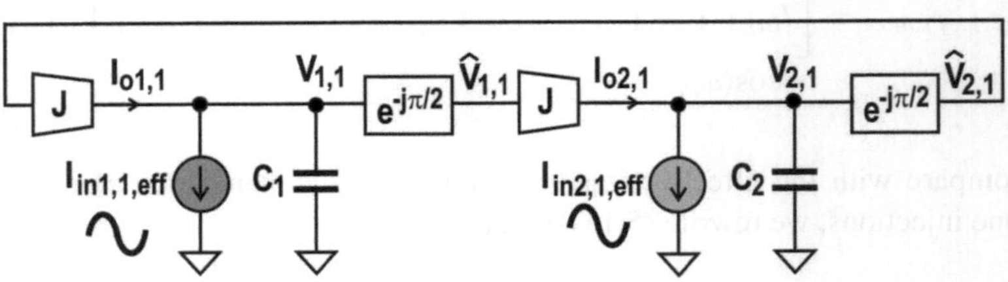

Fig. 5.15 First-order Volterra circuit of dual-comparator relaxation oscillator with dual multitone injections. The injection signals are the effective injection signals at the frequency of the external injection signals ω_{inj}

Following the same approach as that for studying the impact of the degree of nonlinearity on the lock range of the dual-comparator relaxation oscillator given in Sect. 5.2, we obtain the lock range

$$|\Delta\omega| \approx \frac{1}{J}\frac{\omega_o}{2Q(\omega_o)}\frac{1}{|V_{1,1}|} \times \sqrt{|I_{in2,1,eff}|^2 + \frac{|I_{in1,1,eff}|^2}{|H_{o2}|^2} + \frac{e^{j\frac{\pi}{2}}}{H_{o2}} I_{in1,1,eff} I^*_{in2,1,eff} + \frac{e^{-j\frac{\pi}{2}}}{H^*_{o2}} I^*_{in1,1,eff} I_{in2,1,eff}}. \tag{5.71}$$

Making use of

$$H_{o1}(\omega_o + \Delta\omega) \approx H_{o1}(\omega_o) = e^{-j\frac{\pi}{2}}, \tag{5.72a}$$

$$H_{o2}(\omega_o + \Delta\omega) \approx H_{o2}(\omega_o) = e^{-j\frac{\pi}{2}}, \tag{5.72b}$$

which was derived earlier in (5.32) and noting that at the optimal injection phase, the effective injection signals are maximized and their phase difference is 180°, i.e., $I_{in1,1,eff} = -I_{eff1}$ and $I_{in2,1,eff} = I_{eff1}$, we have

$$\begin{aligned} |\Delta\omega| &\approx \frac{1}{J}\frac{\omega_o}{Q(\omega_o)}\frac{I_{eff1}}{|V_{1,1}|} \\ &= \frac{\omega_o}{Q(\omega_o)}\frac{I_{eff1}}{|I_{o2,1}|}. \end{aligned} \tag{5.73}$$

Equation (5.73) is the same as that with unitone injections. Since the effective injection signals of the oscillator with multitone injections is larger than that with unitone injections, the lock range of the oscillator with multitone injections is larger as compared with that with unitone injections.

The preceding results reveal that for the injection-locked pre-scalar of a frequency synthesizer with a ring oscillator, in order to maximize the lock range of the pre-scalar, the injection signal of the pre-scalar should come from the output of the ring oscillator of the synthesizer directly without any filtering so as to preserve all the harmonics of the injection signal. This, in turn, maximizes the lock range of the pre-scalar.

5.5 Single Multitone Injection Versus Dual Multitone Injections

In the previous sections, we investigated the dual-comparator relaxation oscillator with single unitone injection and dual unitone injections. We showed that the lock range of the dual-comparator relaxation oscillator with dual unitone injections is twice that with single unitone injection provided that the injection phases of the dual injections are properly chosen. In this section, we investigate whether a similar result also exists for the dual-comparator relaxation oscillator with single multitone injection and that with dual multitone injections.

Assume the oscillator only has a single multitone injection at node 2 and no injection at node 1, i.e., $i_{inj1} = 0$ and $i_{inj2} \neq 0$. Following the same procedures as those for the oscillator with dual multitone injections, we obtain the effective injections of the first-order Volterra circuit of the oscillator

$$i_{in1,1,eff} = \frac{J}{8}\left(V_{2,1,1,m}^3 + V_{2,1,1,m}^2 V_{2,1,3,m} + V_{2,1,3,m}^2 V_{2,1,5,m}\right)\sin\left(\omega_{inj}t - \frac{\pi}{2}\right), \tag{5.74a}$$

$$\begin{aligned} i_{in2,1,eff} &= I_{inj,1}\sin(\omega_{inj}t + \phi_{inj}) \\ &\quad - \frac{J}{8}\left(V_{1,1,1,m}^3 + V_{1,1,1,m}^2 V_{1,1,3,m} + V_{1,1,3,m}^2 V_{1,1,5,m}\right)\sin\left(\omega_{inj}t - \frac{\pi}{2}\right). \end{aligned} \tag{5.74b}$$

It is seen from (5.74) that similar to the dual-comparator relaxation oscillator with single unitone injection, the third-order Volterra circuit contributes to the effective injection signals of the first-order Volterra at both nodes 1 and 2. The optimal injection phase at which the effective injection signals are maximized is given by

$$\phi_{inj}^* = \frac{\pi}{2}. \tag{5.75}$$

The maximum effective injection signals are given by

$$i_{in1,1,eff,max} = -\frac{J}{4}\left(V_{2,1,1,m}^3 + V_{2,1,1,m}^2 V_{2,1,3,m} + V_{2,1,3,m}^2 V_{2,1,5,m}\right)\cos(\omega_{inj}t), \tag{5.76a}$$

$$\begin{aligned} i_{in2,1,eff,max} &= \left[I_{inj,1} + \frac{J}{4}\left(V_{1,1,1,m}^3 + V_{1,1,1,m}^2 V_{1,1,3,m} + V_{1,1,3,m}^2 V_{1,1,5,m}\right)\right] \\ &\quad \times \cos(\omega_{inj}t). \end{aligned} \tag{5.76b}$$

Following the same approach as that for deriving the lock range of the oscillator with dual multitone injections, one can obtain the lock range of the oscillator with only single multitone injection

$$\begin{aligned} |\Delta\omega| &\approx \frac{1}{J}\frac{\omega_o}{2Q(\omega_o)}\frac{1}{|V_{1,1}|} \\ &\quad \times \sqrt{|I_{in2,1,eff}|^2 + \frac{|I_{in1,1,eff}|^2}{|H_{o2}|^2} + \frac{e^{j\frac{\pi}{2}}}{H_{o2}} I_{in1,1,eff} I_{in2,1,eff}^* + \frac{e^{-j\frac{\pi}{2}}}{H_{o2}^*} I_{in1,1,eff}^* I_{in2,1,eff}}. \end{aligned} \tag{5.77}$$

If the contribution of the third-order Volterra circuit to the effective injection signals of the first-order Volterra circuit is much smaller as compared with the external injection signal, i.e., $I_{in1,1,eff,max} \ll I_{in2,1,eff,max}$, Eq. (5.77) can be simplified to

$$|\Delta\omega| \approx \frac{1}{2}\frac{1}{J}\frac{\omega_o}{Q(\omega_o)}\frac{|I_{inj,1,eff}|}{|V_{1,1}|}. \tag{5.78}$$

A comparison of (5.78) with the lock range with dual multitone injections given by (5.73) reveals that the lock range of the dual-comparator relaxation oscillator with dual multitone injections is twice that with one multitone injection, provided that the two injection signals are injected into the oscillator with proper injection phases, specifically, the injection phase of the current injected into node 2 of the oscillator is $\frac{\pi}{2}$ and the current injected into node 1 of the oscillator is out of phase, i.e., 180° with that injected into node 2.

As an example, consider again the dual-comparator relaxation oscillator shown in Fig. 5.9 [106]. Three cases are investigated:

(a) Double unitone injections:

$$i_{inj1} = I_{inj}\sin(\omega_{inj}t), \tag{5.79a}$$

$$i_{inj2} = -I_{inj}\sin(\omega_{inj}t). \tag{5.79b}$$

(b) Double dual-tone injections:

$$i_{inj1} = I_{inj}\sin(\omega_{inj}t) + \frac{I_{inj}}{3}\sin(3\omega_{inj}t), \tag{5.80a}$$

$$i_{inj2} = -I_{inj}\sin(\omega_{inj}t) - \frac{I_{inj}}{3}\sin(3\omega_{inj}t). \tag{5.80b}$$

(c) Double multitone injections: Two differential square waves of frequency ω_{inj}. The amplitude and duty of the square waves are I_{inj} and 50%, respectively.

Figure 5.16 compares the lock range of the oscillator with double unitone injections, double dual-tone injections, and double multitone injections. The phase difference between the dual injection signals is 180°. It is seen that the lock range of the oscillator with double dual-tone injections is larger as compared with that with double unitone injections. The oscillator with double multitone injections exhibits the largest lock range.

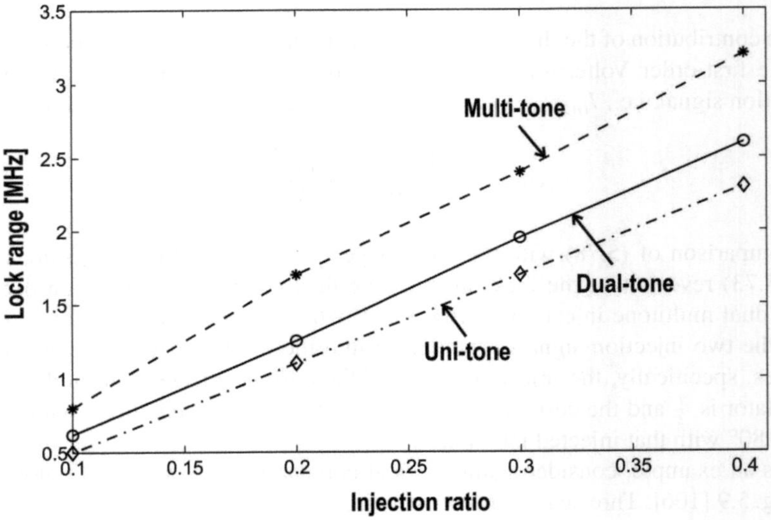

Fig. 5.16 Simulated lock range of dual-comparator relaxation oscillator in Fig. 5.9 with double unitone injections, double dual-tone injections, and double multitone injections. The phase difference between the two injection signals is 180°

5.6 Duty Cycle of Multitone Injection

In the previous sections, we showed that nonharmonic oscillators with multitone injection have a larger lock range as compared with those with unitone injection. The multitone injection signals used to investigate this important characteristic of nonharmonic oscillators are square waves with a 50% duty cycle. As the harmonics of multitone injections contribute to the effective injection signals of the first-order Volterra circuit of the oscillators, the strength of the harmonics of the injection signals should be maximized in order to maximize the effective injection signals of the first-order Volterra circuit subsequently the lock range. In this section, we investigate the impact of the duty cycle of multitone injection signals on the lock range of nonharmonic oscillators, specifically, we want to find out whether there exists an optimal duty cycle of external multitone injection signals that yields the maximum lock range.

Consider the multitone injection signal i_{inj} shown in Fig. 5.17. Since the harmonics of the injection signal is the root cause of the larger lock range of nonharmonic oscillators, we focus on the dependence of the strength of the harmonics of the injection signal on the duty cycle of the injection signal. To find the strength of the harmonics of i_{inj}, we examine its Fourier series expansion

$$i_{inj}(t) = \sum_{k=-\infty}^{\infty} I_{inj,k} e^{jk\omega_{inj}t}, \tag{5.81}$$

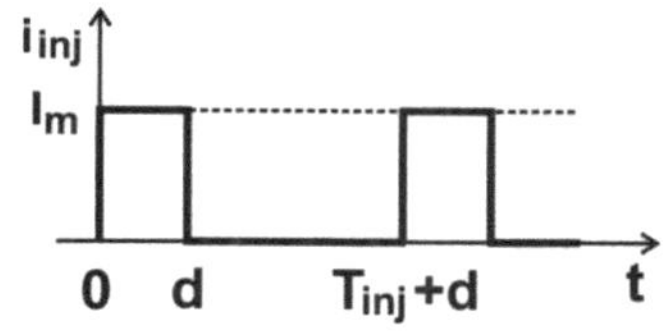

Fig. 5.17 Multitone injection signal

where $\omega_{inj} = 2\pi/T_{inj}$ is the frequency of the injection signal and

$$\begin{aligned} I_{inj,k} &= \frac{1}{T_{inj}} \int_0^{T_{inj}} i_{inj}(t) e^{-jk\omega_{inj}t} dt \\ &= \frac{1}{k\pi} e^{-jk\pi D} \sin(k\pi D), \end{aligned} \tag{5.82}$$

and

$$D = \frac{d}{T_{inj}} \tag{5.83}$$

is the duty cycle of the injection signal. The spectrum of $i_{inj}(t)$ is given by

$$I_{inj}(j\omega) = 2\pi \sum_{k=-\infty}^{\infty} I_{inj,k} \delta(\omega - k\omega_{inj}). \tag{5.84}$$

Equation (5.84) shows that the spectrum of i_{inj} contains a train of impulses at frequencies $k\omega_{inj}$. The strength of the impulses is given by

$$|I_{inj,k}| = \frac{|\sin(k\pi D)|}{k\pi}. \tag{5.85}$$

Let us examine the first five harmonics of the injection signal:

1. $k = 1$: $|I_{inj,1}| = \left|\frac{\sin(\pi D)}{\pi}\right|$, $|I_{inj,1}|$ is minimized when $D = 0, 100\%$ and maximized when $D = 50\%$.
2. $k = 2$: $|I_{inj,2}| = \left|\frac{\sin(2\pi D)}{2\pi}\right|$, $|I_{inj,2}|$ is minimized when $D = 0, 50, 100\%$ and maximized when $D = 25, 75\%$.
3. $k = 3$: $|I_{inj,3}| = \left|\frac{\sin(3\pi D)}{3\pi}\right|$, $|I_{inj,3}|$ is minimized when $D = 0, 33, 66, 100\%$ and maximized when $D = 16.7, 50, 83.3\%$.
4. $k = 4$: $|I_{inj,4}| = \left|\frac{\sin(4\pi D)}{4\pi}\right|$, $|I_{inj,4}|$ is minimized when $D = 0, 25, 50, 75, 100\%$ and maximized when $D = 12.5, 37.5, 62.5, 87.5\%$.
5. $k = 5$: $|I_{inj,5}| = \left|\frac{\sin(5\pi D)}{5\pi}\right|$, $|I_{inj,5}|$ is minimized when $D = 0, 20, 40, 60, 80, 100\%$ and maximized when $D = 10, 30, 50, 70, 90\%$.

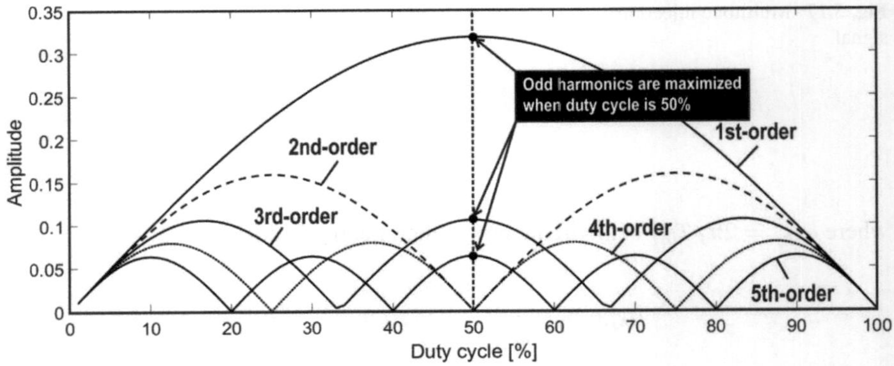

Fig. 5.18 Dependence of the strength of the harmonics of multitone injection signal on its duty cycle

Figure 5.18 plots the dependence of the strength of the harmonics of i_{inj} on the duty cycle. Because the first-harmonic frequency of the injection signal is ω_{inj}, which is also the frequency of the effective injection signal of the first-order Volterra circuit of the oscillator under the injection of i_{inj}, the first harmonic of the injection signal therefore should be maximized by properly choosing the duty cycle of i_{inj}. To achieve this, the duty cycle of the injection signal should be set to 50%. At 50% duty cycle, the second-order and fourth-order harmonics of the injection signal vanish while the third-order and fifth-order harmonics of the injection signal are maximized, thereby strengthening the effective injection signal of the first-order Volterra circuit. The preceding analysis reveals that the optimal duty cycle of the multitone injection signal is 50%.

5.7 Harmonic Attenuation Mechanism

In this section, we compare harmonic oscillators and nonharmonic oscillators in terms of lock range so as to explore the root cause of their distinct lock ranges. As we mentioned before that the dual-comparator relaxation oscillator can be represented by the system shown in Fig. 5.19 with nonlinear transconductors depicting the comparators, and r_{o1} and r_{o2} are the output resistance of transconductors 1 and 2, respectively. The voltage of node 1 and that of node 2 are given by

$$V_1 = \frac{r_{o1}}{sr_{o1}C_1 + 1}(I_{o1} - I_{inj1}), \tag{5.86a}$$

$$V_2 = \frac{r_{o2}}{sr_{o2}C_2 + 1}(I_{o2} - I_{inj2}). \tag{5.86b}$$

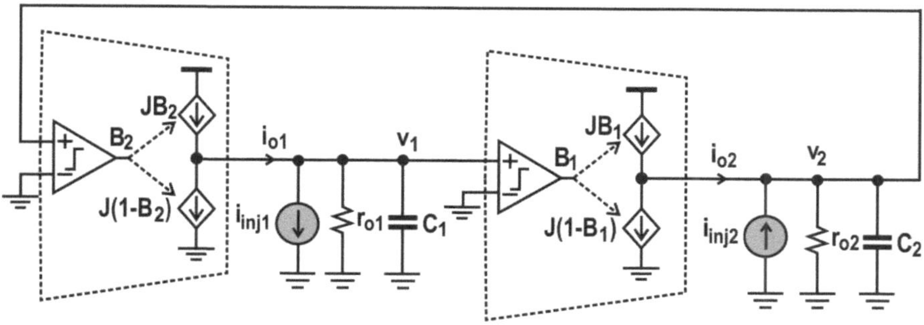

Fig. 5.19 Block diagram of dual-comparator relaxation oscillators

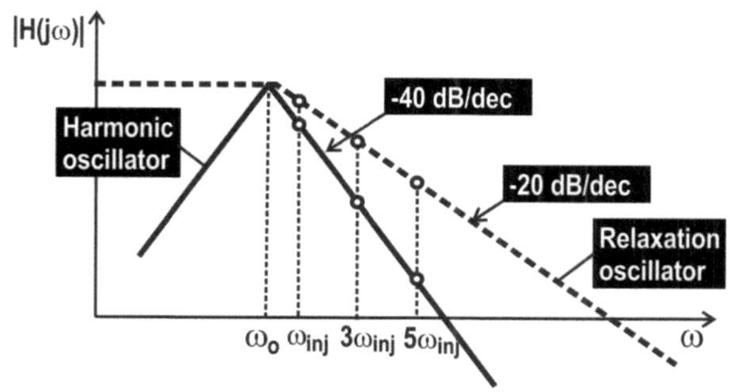

Fig. 5.20 Attenuation of the frequency components of the output current of the nonlinear transconductors in relaxation oscillators and harmonic oscillators

Since the output current of the nonlinear transconductors contains frequency components at ω_{inj}, $3\omega_{inj}$, and $5\omega_{inj}$, these frequency components are shaped by the low-pass mechanism formed by the output resistance of the transconductors and the integrating capacitors with -3 dB frequency given by

$$\omega_{-3dB} = \frac{1}{r_{o1,2}C_{1,2}} \tag{5.87}$$

and a -20 dB/dec rate of attenuation in the stop band, as shown in Fig. 5.20. As the contribution of the third-order Volterra circuit to the effective injection signals of the first-order Volterra circuit is proportional to $v_{n,1}^3$, $n = 1, 2$, the less the attenuation of $v_{n,1}$ by the low-pass mechanism, the larger the effective injection signals of the first-order Volterra circuit and subsequently the larger the lock range of the oscillator.

Let us now consider the harmonic oscillator in Fig. 5.21 with its block diagram also shown in the figure. The nonlinear transconductors are used to depict the MOS transistors. Note that since MOS transistors operate in saturation during state transition, the nonlinear characteristics of the transconductors are significantly milder as compared with those of the comparators of the relaxation oscillator. Another distinct characteristic of the harmonic oscillator is the presence of inductors that are in parallel with the capacitors. It can be shown that

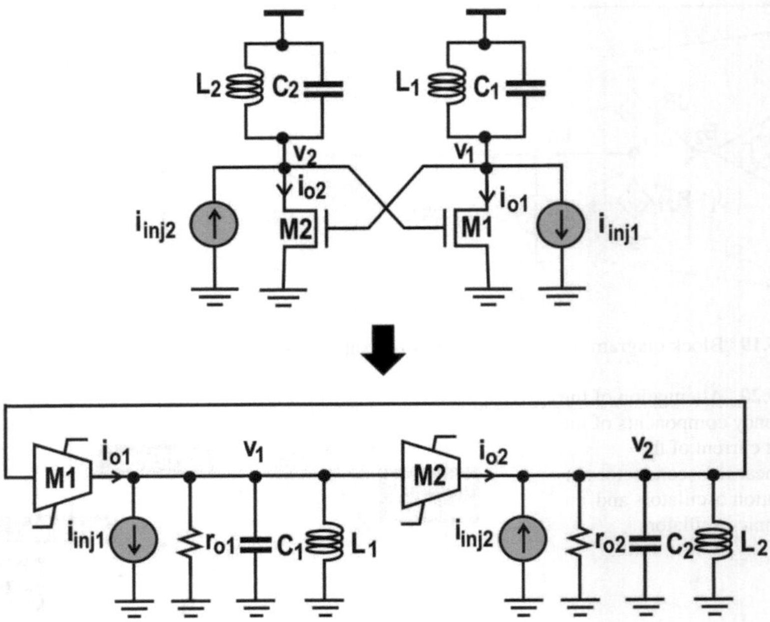

Fig. 5.21 Block diagram of harmonic oscillators

$$V_1 = \frac{1}{C_1} \frac{s(I_{o1} - I_{inj1})}{s^2 + \frac{s}{r_{o1}C_1} + \frac{1}{L_1C_1}}, \tag{5.88a}$$

$$V_2 = \frac{1}{C_2} \frac{s(I_{o2} + I_{inj2})}{s^2 + \frac{s}{r_{o2}C_2} + \frac{1}{L_2C_2}}. \tag{5.88b}$$

It is evident from (5.38) that V_1 and V_2 have a band-pass characteristic with the center frequency of the pass-band given by

$$\omega_o = \frac{1}{\sqrt{L_{1,2}C_{1,2}}}. \tag{5.89}$$

The magnitude of V_1 and V_2 levels off from the center frequency of the passband at a rate of -40 dB/dec when frequency deviates from ω_o. The frequency components of the output current of the transconductors at ω_{inj}, $3\omega_{inj}$, and $5\omega_{inj}$ are now attenuated more severe as compared with the dual-comparator relaxation oscillator studied earlier. As the contribution of the third-order Volterra circuit to the effective injection signals of the first-order Volterra circuit is proportional to $v_{n,1}^3$, $n = 1, 2$, the more severe the attenuation of $v_{n,1}$ by the bandpass mechanism, the smaller the effective injection signals of the first-order Volterra circuit and subsequently the smaller the lock range of the oscillator. It should also be noted that the higher the quality factor of the LC oscillator, the more attenuation of the higher-order frequency components of the output current of the transconductors, the weaker the injection signals and subsequently the smaller the lock range.

In addition to the attenuation of the harmonic components of the output of the transconductors, the degree of the nonlinearity of the transconductors in the dual-comparator relaxation oscillator is higher as compared with that of the transconductors in the ring oscillator. As a result, the amplitude of the harmonic components of the output of the transconductors in the relaxation oscillator is higher as compared with that in the ring oscillator. We therefore conclude from the preceding analysis that the effective injection signals of the relaxation oscillator is larger as compared with those of the harmonic oscillator. As a result, the relaxation oscillator exhibits a larger lock range than that of the harmonic oscillator.

5.8 Chapter Summary

This chapter investigated several techniques that increase the lock range of non-harmonic oscillators. The impact of the degree of the nonlinear characteristics of nonharmonic oscillators on the lock range of the oscillators was explored. We showed that the higher the degree of the nonlinear characteristics of nonharmonic oscillators, the more the high-order Volterra circuits contribute to the effective injection signals of the first-order Volterra circuit of the oscillators and the larger the lock range of the oscillators.

The effective injection signals of the first-order Volterra circuit of nonharmonic oscillators with single unitone injection and that with dual unitone injections were compared. We showed that high-order Volterra circuits contribute to the effective injection signals of the first-order Volterra circuit even though only one external injection exists. The lock range of nonharmonic oscillators with dual unitone injections is twice that with single unitone injection.

We further investigated nonharmonic oscillators with single multitone injection and dual multitone injections. We showed that the high-order Volterra circuit of nonharmonic oscillators with multitone injection contribute more to the effective injection signals of the first-order Volterra circuit as compared with those with unitone injection, thereby yielding a larger lock range. We further showed that the high-order Volterra circuits of nonharmonic oscillators contribute to the effective injection signals of the first-order Volterra circuit of the oscillators even only one external multitone injection exists. The lock range of nonharmonic oscillators with dual multitone injections is twice that with single multitone injection.

The impact of the duty cycle of multitone injection signals on the lock range of nonharmonic oscillators was investigated. We showed that when the duty cycle of a multitone injection signal is 50%, its odd harmonics are maximized while their even harmonics vanish, thereby yielding the maximum effective injection signals of the first-order Volterra circuit of the oscillators and subsequently the largest lock range.

Finally, the impact of frequency attenuation mechanisms on the lock range of nonharmonic and harmonic oscillators was compared. We showed that the existence of the second-order frequency attenuation mechanism in harmonic oscillators effectively lowers the harmonics of the output of the oscillators and subsequently the effective injection signals of their first-order Volterra circuit. For nonharmonic oscillators, the frequency attenuation mechanism is first-order. As a result, the harmonic components of the output of these oscillators are less attenuated. This is echoed with larger effective injection signals of the first-order Volterra circuit of the oscillators and subsequently a larger lock range.

References

1. R. Adler, A study of locking phenomena in oscillators. Proc. Inst. Radio Eng. **34**(6), 351–357 (1946)
2. E. Appleton, The automatic synchronization of triode oscillators. Proc. Camb. Philos. Soc. **21**, 231–248 (1922–1923)
3. J. Bae, L. Yan, H. Yoo, A low energy injection-locked FSK transceiver with frequency-to-amplitude conversion for body sensor applications. IEEE J. Solid State Circuits **46**(4), 928–937 (2011)
4. J. Bae, H. Yoo, A 45 μW injection-locked FSK wake-up receiver with frequency-to-envelope conversion for crystal-less wireless body area network. IEEE J. Solid State Circuits **50**(6), 1351–1360 (2015)
5. G. Balamurugan, N. Shanbhag, Modeling and mitigation of jitter in multi Gbps source-synchronous I/O links, in *Proceedings of the International Conference on Computer Design* (IEEE, Piscataway, 2003), pp. 254–260
6. G. Beers, A frequency-dividing locked-in oscillator frequency-modulation receiver. Proc. Inst. Radio Eng. **32**(12), 730–737 (1944)
7. G. Begemann, A. Jacob, Conversion gain of MESFET drain mixers. IEEE Lett. **15**(18), 567–568 (1979)
8. A. Buonomo, A. Lo Schiavo, M. Awan, M. Asghar, M. Kennedy, A CMOS injection-locked frequency divider optimized for divide-by-two and divide-by-three operation. IEEE Trans. Circuits Syst. I **60**(12), 3126–3135 (2013)
9. B. Casper, F. O'Mahony, Clocking analysis, implementation and measurement techniques for high-speed data links—a tutorial. IEEE Trans. Circuits Syst. I **56**(1), 17–39 (2009)
10. J. Cayrou, M. Gayral, J. Graffeuil, J. Sautereau, Simple expression for conversion gain of MESFET drain mixers. IEEE Lett. **29**(17), 1514–1516 (1993)
11. H. Chen, D. Chang, Y. Juang, S. Lu, A 30-GHz wideband low-power CMOS injection-locked frequency divider for 60-GHz wireless-LAN. IEEE Microwave Wireless Lett **18**(2), 145–147 (2008)
12. J. Chen, A. Hu, Y. Fan, R. Bashirullah, Noise suppression in injection-locked ring oscillators. IEEE Lett. **48**(6), 323–324 (2012)
13. D. Cherniak, R. Nonis, F. Padovan, A precision 140 MHz relaxation oscillator in 40 nm CMOS with 28 ppm/°C frequency stability for automotive SoC applications, in *Proceedings of the IEEE Radio Frequency Integrated Circuits Symposium* (IEEE, Piscataway, 2017), pp. 57–60
14. J. Chien, L. Lu, Analysis and design of wideband injection-locked ring oscillators with multiple-input injection. IEEE J. Solid State Circuits **42**(9), 1906–1915 (2007)

F. Yuan, *Injection-Locking in Mixed-Mode Signal Processing*,
https://doi.org/10.1007/978-3-030-17364-7

15. J. Chien, L. Lu, 40 GHz wide-locking-range regenerative frequency divider and low-phase-noise balanced VCO in 0.18 μm CMOS, in *IEEE International Solid-State Circuits Conference—Digest of Technical Papers* (IEEE, Piscataway, 2007), pp. 544–621
16. S. Cho, S. Kim, M. Choo, J. Lee, H. Ko, S. Jang, S. Chu, W. Bae, Y. Kim, D. Jeong, A 5-GHz subharmonically injection-locked all-digital PLL with complementary switched injection, in *Proceedings of the European Solid-State Circuits Conference* (IEEE, Piscataway, 2015), pp. 384–387
17. M. Choo, H. Ko, S. Cho, K. Lee, D. Jeong, An optimum injection-timing tracking loop for 5-GHz, 1.13-mW/GHz RO-based injection-locked PLL with 152-fs integrated jitter. IEEE Trans. Circuits Syst. II **65**(12), 1819–1823 (2018)
18. Y. Chuang, S. Lee, S. Jang, J. Chao, M. Juang, A ring-oscillator-based wide locking range frequency divider. IEEE Microwave Wireless Compon. Lett. **16**(8), 470–472 (2006)
19. F. Ellinger, L. Rodoni, G. Sialm, C. Kromer, G. von Buren, M. Schmatz, C. Menolfi, T. Toifl, T. Morf, M. Kossel, H. Jackel. 30–40-GHz drain-pumped passive-mixer MMIC fabricated on VLSI SOI CMOS technology. IEEE Trans. Microwave Theory Tech. **52**(5), 1382–1391 (2004)
20. EPC™, *Radio-Frequency Identity Protocols Class-1 Generation-2 UHF RFID Protocol for Communications at 860 MHz–960 MHz, Version 1.0.9* (EPCglobal, 2005)
21. D. Fraser, Synchrobization of oscillators by periodically interrupted waves. Proc. Inst. Radio Eng. **45**(9), 1256–1268 (1957)
22. M. Fujishima, K. Amamoto, A 1.0 V 10.2 GHz CMOS frequency divider with differential injection locking, in *Proceedings of the IEEE Topical Conference on Wireless Communications Technology* (IEEE, Piscataway, 2003), pp. 164–165
23. P. Gray, P. Hust, S. Lewis, R. Meyer, *Analysis and Design of Analog Integrated Circuits*, 4th edn. (Wiley, New York, 2001)
24. W. Grollitsch, R. Nonis, A fractional-N, all-digital injection-locked PLL with wide tuning range digitally controlled ring oscillator and bang-bang phase detection for temperature tracking in 40 nm CMOS, in *Proceedings of the European Solid-State Circuits Conference* (IEEE, Piscataway, 2016), pp. 201–204
25. A. Hajimiri, S. Limotyrakis, T. Lee, Jitter and phase noise in ring oscillators. IEEE J. Solid State Circuits **34**(6), 790–804 (1999)
26. F. Herzal, B. Razavi, A study of oscillator jitter due to supply and substrate noise. IEEE Trans. Circuits Syst. II **46**(1), 56–62 (1999)
27. M. Hossain, A. Carusone, 7.4 Gb/s 6.8 mW source synchronous receiver in 65 nm CMOS. IEEE J. Solid State Circuits **46**(6), 1337–1348 (2011)
28. K. Hu, R. Bai, T. Jiang, C. Ma, A. Ragab, S. Palermo, P. Chiang, 0.16–0.25 pJ/bit, 8 Gb/s near-threshold serial link receiver with super-harmonic injection-locking. IEEE J. Solid State Circuits **47**(8), 1842–1853 (2012)
29. K. Hu, T. Jiang, J. Wang, F. O'Mahony, P. Chiang, A 0.6 m W/Gb/s, 6.4–7.2 Gb/s serial link receiver using local injection-locked ring oscillators in 90 nm CMOS. IEEE J. Solid State Circuits **45**(4), 899–908 (2010)
30. J. Hu, B. Otis, A 3 μW, 400 MHz divide-by-5 injection-locked frequency divider with 56% lock range in 90 nm CMOS, in *2008 IEEE Radio Frequency Integrated Circuits Symposium* (IEEE, Piscataway, 2008), pp. 665–668
31. R. Huntoon, A. Weiss, Synchronization of oscillators. Proc. Inst. Radio Eng. **34**(12), 1415–1423 (1947)
32. S. Jang, C. Chang, A 90 nm CMOS LC-Tank divide-by-3 injection-locked frequency divider with record locking range. IEEE Microwave Wireless Compon. Lett. **20**(4), 229–231 (2010)
33. S. Jang, C. Chang, W. Cheng, C. Lee, M. Juang, Low-power divide-by-3 injection-locked frequency dividers implemented with injection transformers. IET Electron. Lett. **45**(5), 240–241 (2009)
34. S. Jang, P. Lu, M. Juang, Divide-by-3 LC injection locked frequency divider with a transformer as an injector's load. Microw. Opt. Technol. Lett. **50**(10), 2722–2725 (2008)

35. J. Jin, X. Yu, J. Zhou, T. Yan, Gigahertz range injection locked frequency dividers with bandwidth enhancement and supply rejection. Electron. Lett. **44**(17), 999–1000 (2008)
36. U. Karthaus, M. Fischer, Fully integrated passive UHF RFID transponder IC with 16.7 μW minimum RF input power. IEEE J. Solid State Circuits **38**(10), 1602–1608 (2003)
37. F. Kocer, M. Flynn, A new transponder architecture with on-chip ADC for long-range telemetry applications. IEEE J. Solid State Circuits **41**(5), 1142–1148 (2006)
38. Y. Lam, S. Kim, A 16.6 μW 32.8 MHz monolithic CMOS relaxation oscillator, in *Proceedings of the IEEE Asian Solid-State Circuits Conference* (IEEE, Piscataway, 2014), pp. 161–164
39. S. Lee, S. Amakawa, N. Ishihara, K. Masu, Low-phase-noise wide-frequency-range ring-VCO-based scalable PLL with subharmonic injection locking in 0.18 μm CMOS, in *Proceedings of the IEEE MTT-S International Microwave Symposium* (IEEE, Piscataway, 2010), pp. 1178–1181
40. M. Lee, W. Dally, T. Greer, H. Ng, R. Farjad-Rad, J. Poulton, R. Senthinathan, Jitter transfer characteristics of delay-locked loops—theories and design techniques. IEEE J. Solid State Circuits **38**(4), 614–621 (2003)
41. K. Lee, S. Kim, Y. Shin, D. Jeong, G. Lim, B. Kim, V. Da Costa, D. Lee, A jitter-tolerant 4.5 Gb/s CMOS interconnect for digital display, in *1998 IEEE International Solid-State Circuits Conference. Digest of Technical Papers, ISSCC* (IEEE, Piscataway, 1998), pp. 310–311
42. D. Lee, T. Lee, Y. Kim, Y. Kim, L. Kim, An injection locked PLL for power supply variation robustness using negative phase shift phenomenon of injection locked frequency divider, in *Proceedings of the IEEE Custom Integrated Circuits Conference* (IEEE, Piscataway, 2015), pp.1–4
43. J. Lee, H. Wang, Study of subharmonically injection-locked PLLs. IEEE J. Solid State Circuits **44**(5), 1539–1553 (2009)
44. Y. Lo, H. Chen, J. Silva-Martinez, S. Hoyos, A 1.8V, sub-mW, over 100% locking range, divide-by-3 and 7 complementary-injection-locked 4 GHz frequency divider, in *Proceedings of IEEE Custom Integrated Circuits Conference* (IEEE, Piscataway, 2009), pp. 259–262
45. T. Lou, Y. Chen, A 0.8-mW 55-GHz dual-injection-locked CMOS frequency divider. IEEE Trans. Microwave Theory Tech. **56**(3), 620–625 (2008)
46. L. Lin, L. Tee, P. Gray, A 1.4 GHz differential low-noise CMOS frequency synthesizer using a wideband PLL architecture, in *2000 IEEE International Solid-State Circuits Conference. Digest of Technical Papers* (IEEE, Piscataway, 2000), pp. 204–205
47. S. Liu, Y. Zheng, W. M. Lim, W. Yang, Ring oscillator based injection locked frequency divider using dual injection paths. IEEE Microwave Wireless Compon. Lett. **25**(5), 322–324 (2015)
48. L. Lu, J. Chien, A wide-band CMOS injection-locked ring oscillator. IEEE Microwave Wireless Compon. Lett. **15**(10):676–678 (2005)
49. R. Mackey, Injection locking of klystron oscillators. IRE Trans. Microwave Theory Tech. **10**(4), 228–235 (1962)
50. B. Mesgarzadeh, A. Alvandpour, A study of injection locking in ring oscillators, in *Proceedings of IEEE International Symposium on Circuits and Systems* (IEEE, Piscataway, 2005), pp. 5465–5468
51. B. Mesgarzadeh, A. Alvandpour, First-harmonic injection-locked ring oscillators, in *Proceedings of IEEE Custom Integrated Circuits Conference* (IEEE, Piscataway, 2006), pp. 733–736,
52. J. Mikulic, G. Schatzberger, A. Baric, A 1-MHz on-chip relaxation oscillator with comparator delay cancellation, in *Proceedings of the IEEE European Solid State Circuits Conference* (IEEE, Piscataway, 2017), pp. 95–98
53. R. Miller, Fractional-frequency generators utilizing regenerative modulation. Proc. Inst. Radio Eng. **27**(7), 446–457 (1939)
54. A. Mirzaei, M. Heidari, A. Abidi, Analysis of oscillators locked by large injection signals: generalized Adler's equation and geometrical interpretation, in *IEEE Custom Integrated Circuits Conference 2006* (IEEE, Piscataway, 2006), pp. 737–740

55. A. Mirzaei, M. Heidari, R. Bagheri, A. Abidi, Multi-phase injection widens lock range of ring-oscillator-based frequency dividers. IEEE J. Solid State Circuits **43**(3), 656–671 (2008)
56. A. Mirzaei, M. Heidari, R. Bagheri, S. Chehrazi, A. Abidi, The quadrature LC oscillator: a complete portrait based on injection locking. IEEE J. Solid State Circuits **42**(9), 1916–1932 (2007)
57. H. Moyer, A. Daryoush, A unified analytical model and experimental validations of injection-locked processes. IEEE Trans. Microwave Theory Tech. **48**(4), 493–499 (2000)
58. A. Musa, K. Okada, A. Matsuzawa, Progressive mixing technique to widen the locking range of high division-ratio injection-locked frequency dividers. IEEE Trans. Microwave Theory Tech. **61**(3), 1161–1173 (2013)
59. H. Ng, R. Farjad-Rad, M. Lee, W. Dally, T. Greer, J. Poulton, J. Edmondson, R. Rathi, R. Senthinathan, A second-order semidigital clock recovery circuit based on injection locking. IEEE J. Solid State Circuits **38**(12), 2101–2110 (2003)
60. F. O'Mahony, S. Shekhar, M. Mansuri, G. Balamurugan, J. Jaussi, J. Kennedy, B. Casper, D. J. Allstot, R. Mooney, A 27 Gb/s forwarded-clock I/O receiver using an injection-locked LC-DCO in 45 nm CMOS, in *2008 IEEE International Solid-State Circuits Conference. Digest of Technical Papers* (IEEE, Piscataway, 2008), pp. 452–627
61. L. Paciorek, Injection locking of oscillators. Proc. IEEE **53**(11), 1723–1727 (1965)
62. F. Plessas, F. Gioulekas, G. Kalivas, Phase noise performance of fully differential sub-harmonic injection-locked PLL. Electron. Lett. **46**(19), 1319–1321 (2010)
63. H. Rategh, T. Lee, Superharmonic injection-locked frequency dividers. IEEE J. Solid State Circuits **34**(6), 813–821 (1999)
64. B. Razavi, A study of phase noise in CMOS oscillators. IEEE J. Solid State Circuits **31**(3), 331–343 (1996)
65. B. Razavi, *Design of Integrated Circuits for Optical Communications* (McGraw-Hill, Boston, 2003)
66. B. Razavi, A study of injection locking and pulling in oscillators. IEEE J. Solid State Circuits **39**(9), 1415–1424 (2004)
67. M. Schetzen, *The Volterra and Wiener Theory of Nonlinear Systems* (Wiley, New York, 1981)
68. I. Schmideg, Harmonic synchronization of nonlinear oscillators. Proc. IEEE **59**(8), 1250–1251 (1971)
69. A. Sedra, K. Smith, *Microelectronics Circuits* (Oxford University Press, London, 1998)
70. S. Shekhar, M. Mansuri, F. O'Mahony, G. Balamurugan, J. E. Jaussi, J. Kennedy, D. Allstot, R. Mooney, B. Casper, Strong injection locking in low-Q LC oscillators: modeling and application in a forwarded-clock I/O receiver. IEEE Trans. Circuits Syst. I **56**(8), 1818–1829 (2009)
71. H. Shin, H. Kim, Extraction technique of differential second harmonic output in CMOS LC VCO. IEEE Microwave Wireless Compon. Lett. **17**(5), 379–381 (2007)
72. D. Shin, S. Park, S. Raman, K. Koh, A subharmonically injection-locked PLL with 130 fs RMS jitter at 24 GHz using synchronous reference pulse injection from nonlinear VCO envelope feedback, in *Proceedings of the IEEE Radio Frequency Integrated Circuits Symposium* (IEEE, Piscataway, 2017), pp . 100–103
73. N. Soltani, F. Yuan, Non-harmonic injection-locked phase-locked loops with applications in remote frequency calibration of passive wireless transponders. IEEE Trans. Circuits Syst. I **57**(12), 2381–2393 (2010)
74. M. Tiebout, A CMOS direct injection-locked oscillator topology as high-frequency low-power frequency divider. IEEE J. Solid State Circuits **39**(7), 1170–1174 (2004)
75. A. Tofangdarzade, A. Jalali, An efficient method to analyze lock range in ring oscillators with multiple injections. IEEE Trans. Circuits Syst. II **62**(11), 1013–1017 (2015)
76. A. Tofangdarzade, A. Tofangdarzade, N. Saniei, Strong injection locking and pulling in LC multiphase oscillators with multiple injection signals. IEEE Trans. Circuits Syst. II (2018, accepted)
77. H. Tong, S. Cheng, A. Karsilayan, J. Martinez, An injection-locked frequency divider with multiple highly nonlinear injection stages and large division ratios. IEEE Trans. Circuits Syst. II **54**(4), 313–317 (2007)

78. P. Tsai, C. Liu, T. Huang, Wideband injection-locked divide-by-3 frequency divider design with regenerative second-harmonic feedback technique, in *Proceedings of the European Microwave Integrated Circuit Conference* (IEEE, Piscataway, 2012), pp. 293–296
79. D. Tucker, Forced oscillations in oscillator circuits, and the synchronization of oscillators. J. Inst. Radio Eng. Part III: Radio Commun. Eng. **92**(19), 226–234 (1945)
80. D. Tucker, Forced oscillations in oscillator circuits, and the synchronization of oscillators. J. Inst. Radio Eng. Part I: Gen. **93**(61), 57–58 (1946)
81. B. van der Pol, The nonlinear theory of electric oscillations. Proc. Inst. Radio Eng. **22**(9), 1051–1086 (1934)
82. C. van den Bos, C. Verhoeven, Frequency division using an injection-locked relaxation oscillator, in *Proceedings of IEEE International Symposium on Circuits and Systems*, vol. 4 (IEEE, Piscataway, 2002), pp. 517–520
83. S. Verma, H. Rategh, T. Lee, A unified model for injection-locked frequency divider. IEEE J. Solid State Circuits **38**(6), 1015–1027 (2003)
84. J. Vincent, On some experiments in which two neighbouring maintained oscillatory circuits affect a resonating circuit. Proc. Phys. Soc. Lond. **32**, 84–91 (1919)
85. J. Vlach, K. Singhal, *Computer Methods for Circuit Analysis and Design* (Van Nostrand Reinhold, New York, 1993)
86. P. Wampacq, W. Sansen, *Distortion Analysis of Analog Integrated Circuits* (Kluwer Academic, Boston, 1998)
87. Y. Wan, X. Lai, J. Roychowdhury, Understanding injection locking in negative-resistance LC oscillators intuitively using nonlinear feedback analysis, in *Proceedings of IEEE Custom Integrated Circuits Conference* (IEEE, Piscataway, 2005), pp. 729–732
88. H. Wang, A 1.8 V 3 mW 16.8 GHz frequency divider in 0.25 μm CMOS, in *IEEE International Solid-State Circuits Conference. Digest of Technical Papers* (IEEE, Piscataway, 2000), pp. 196–197
89. J. Wang, W. L. Goh, X. Liu, J. Zhou, A 12.77-MHz 31 ppm/°C on-chip RC relaxation oscillator with digital compensation technique. IEEE Trans. Circuits Syst. I **63**(11), 1816–1824 (2016)
90. C. Wei, T. Kuan, S. Liu, A subharmonically injection-locked PLL with calibrated injection pulsewidth. IEEE Trans. Circuits Syst. II **62**(6), 548–552 (2015)
91. H. Wu, A. Hajimiri, A 19 GHz 0.5 mW 0.35 μm CMOS frequency divider with shunt-peaking locking-range enhancement, in *IEEE International Solid-State Circuits Conference. Digest of Technical Papers* (IEEE, Piscataway, 2005), pp. 412–413
92. V. Wu, C. Yu, Design and analysis of a millimeter-wave direct injection-locked frequency divider with large frequency locking range. IEEE Trans. Microwave Theory Tech. **55**(8), 1649–1658 (2007)
93. H. Wu, L. Zhang, A 16-to-18 GHz 0.18-m Epi-CMOS divide-by-3 injection-locked frequency divider, in *IEEE International Solid-State Circuits Conference—Digest of Technical Papers* (IEEE, Piscataway, 2006), pp. 2482–2491
94. K. Yamamoto, M. Fujishima, 55 GHz CMOS frequency divider with 3.2 GHz locking range, in *Proceedings of the 30th European Solid-State Circuits Conference* (IEEE, Piscataway, 2004), pp. 135–138
95. K. Yamamoto, M. Fujishima, A 44-μW 4.3-GHz injection-locked frequency divider with 2.3-GHz locking range. IEEE J. Solid State Circuits **40**(3), 671–677 (2005)
96. K. Yamamoto, M. Fujishima, 70 GHz CMOS harmonic injection-locked divider, in *IEEE International Solid-State Circuits Conference—Digest of Technical Papers* (IEEE, Piscataway, 2006), pp. 2472–2473
97. Y. Yeh, H. Chang, Design and analysis of a W-band divide-by-three injection-locked frequency divider using second harmonic enhancement technique. IEEE Trans. Microwave Theory Tech. **60**(6), 1617–1625 (2012)
98. X. Yi, C. Boon, M. Do, K. Yeo, W. Lim, Design of ring-oscillator-based injection-locked frequency dividers with single-phase inputs. IEEE Microwave Wireless Compon. Lett. **21**(10), 559–561 (2011)

99. X. Yu, H. Cheema, R. Mahmoudi, A. van Roermund, X. Yan, A 3 mW 54.6 GHz divide-by-3 injection locked frequency divider with resistive harmonic enhancement. IEEE Microwave Wireless Compon. Lett. **19**(9), 575–577 (2009)
100. X. Yu, M. Do, J. Ma, W. Lim, K. Yeo, X. Yan, Sub-1 V low power wide range injection-locked frequency divider. IEEE Microwave Wireless Compon. Lett. **17**(7):528–530 (2007)
101. F. Yuan, *Computer Methods for Analysis of Mixed-Mode Switching Circuits* (Kluwer Academic, New York, 2004)
102. F. Yuan, *CMOS Active Inductors and Transformers* (Springer, New York, 2008)
103. F. Yuan, M. Li, A new CMOS class AB serial link transmitter with low supply voltage sensitivity. Analog Integr. Circ. Sig. Process **49**(2), 171–180 (2006)
104. F. Yuan, N. Soltani, A low-voltage low VDD sensitivity relaxation oscillator for passive wireless microsystems. IET Electron. Lett. **45**(21), 1057–1058 (2009)
105. F. Yuan, Y. Zhou, A phasor-domain study of lock range of harmonic oscillators with multiple injections. IEEE Trans. Circuits Syst. II **59**(8), 466–470 (2012)
106. F. Yuan, Y. Zhou, Frequency-domain study of lock range of non-harmonic oscillators with multiple multi-tone injections. IEEE Trans. Circuits Syst. I **60**(6), 1395–1406 (2013)
107. F. Yuan, Y. Zhou, Injection signaling in relaxation oscillators. Analog Integr. Circ. Sig. Process **2019**, 1–6 (2019)
108. L. Zhang, B. Ciftcioglu, M. Huang, H. Wu, Injection-locked clocking: a new GHz clock distribution scheme, in *Proceedings of the IEEE Custom Integrated Circuits Conference* (IEEE, Piscataway, 2006), pp. 785–788
109. Z. Zhang, L. Liu, N. Wu, A novel 2.4-to-3.6 GHz wideband subharmonically injection-locked PLL with adaptively-aligned injection timing, in *Proceedings of the IEEE Asian Solid-State Circuits Conference* (IEEE, Piscataway, 2014), pp. 369–372
110. X. Zhang, X. Zhou, A. Daryoush, A theoretical and experimental study of the noise behavior of subharmonically injection locked local oscillators. IEEE Trans. Microwave Theory Tech. **40**(5), 895–902 (1992)
111. Y. Zhou, F. Yuan, Subthreshold CMOS active inductors with applications to low-power injection-locked oscillators for passive wireless microsystems, in *Proceedings of the Midwest Symposium on Circuits and Systems* (IEEE, Piscataway, 2010), pp. 885–888
112. Y. Zhou, F. Yuan. A study of lock range of injection-locked CMOS active-inductor oscillators using a linear control system approach. IEEE Trans. Circuits Syst. II **58**(10), 627–631 (2011)
113. Y. Zhou, F. Yuan, Study of injection-locked non-harmonic oscillators using volterra series. IET Circuits Devices and Syst. **9**(2), 119–130 (2015)

Index

F. Yuan, *Injection-Locking in Mixed-Mode Signal Processing*,
https://doi.org/10.1007/978-3-030-17364-7

MIX
Papier aus verantwortungsvollen Quellen
Paper from responsible sources
FSC® C105338

If you have any concerns about our products,
you can contact us on
ProductSafety@springernature.com

In case Publisher is established outside the EU,
the EU authorized representative is:
Springer Nature Customer Service Center GmbH
Europaplatz 3, 69115 Heidelberg, Germany

Printed by Libri Plureos GmbH
in Hamburg, Germany